Jobs You Can Live With
Working at the Crossroads of Science, Technology, and Society

Fifth Edition

Managing Editor: Susan M. Higman
Contributing Editors: Gregory Craven and Lawrence Joseloff

Student Pugwash USA
815 15th Street, NW, Suite 814, Washington, DC 20005

Cover and text design: Public Interest GRFX, Philadelphia, PA
Printing: Bang Printing, Brainerd, MN

Copyright ©1996, Student Pugwash USA

All rights reserved. No part of this book may be reproduced or transmitted in any form or by any means, electronic or mechanical, including photocopying, recording, or any information storage and retrieval system without written permission from Student Pugwash USA.

ISBN 0-9639007-1-4
Library of Congress Catalog Card Number: 96-67359

Printed on recycled paper with soy ink.

Dedication

This book is dedicated to all those who are committed to working for a better future. In particular, we acknowledge the Pugwash Conferences on Science and World Affairs for their years of peace-seeking work and their inspiration to the next generation. We also commend the students and young professionals who are beginning their careers with the goal of creating a more just, secure, and sustainable world.

Acknowledgments

This directory would not be possible without the efforts of many people. Thanks to all who have given their time, efforts, and creativity to this project. Thanks to the person at each organization tasked with filling out the profile form. Thanks also to those who helped bring the book to production, especially: Fran Briney for proofreading the manuscript, Ennis Carter at Public Interest GRFX, and Debbie Roberts at Bang Printing.

In particular, thanks to the team here at Student Pugwash USA, both past and present. Editors of prior editions offered words of wisdom along the way, including Rachel Lostumbo and Barry Lasky. Contributing editors Greg Craven and Larry Joseloff brought the 5th edition from idea to reality. Other Student Pugwash USA staff and interns gave their support throughout the process during 1995-96: David Andersen (current National Chapter Coordinator), Laura Baraff, Laura Blasi, Jennifer Chen, Marianne Doyle, Betsy Fader (former Executive Director), Lauren Ginns, Andrea Hovanec, Sandra Ionno (current Executive Director), Sean Kennedy, Constance Lassiter, Matthew Lee, Beth McKay, Nicola Short, Russell Singleton, Amanda Smith, and Nick White.

Past and present members of the board of directors of Student Pugwash USA offered guidance and support. Current board members: Edward Bales, Michael Berger, Richard Bryant, Jr., Anne Cahn, Peter Carpenter, Rebecca Derrig-Green, Mohammed El-Ashry, Paul Jellinek, Alan McGowan, Ann Moore, Bob Murching, Indira Nair, Constance Pechura (chair), Dann Sklarew, Anna Yusim. Recent board members: Ruth Adams, Taft Broome, Rajesh Krishnan, Nicholas Steneck (former chair).

Special thanks to those who have recently contributed to Student Pugwash USA, making our work possible: Apple Computer, Inc., Carpenter Family Trust, Ciba Educational Foundation, Ciba Limited, The Commonwealth Fund, Department of Energy (US), Cyrus Eaton Foundation, W. Alton Jones Foundation, Henry P. Kendall Foundation, Jeffrey Leifer (founder), John D. and Catherine T. MacArthur Foundation, Stewart R. Mott Charitable Trust, National Science Foundation (US), New-Land Foundation, Ploughshares Fund, Rockefeller Family Associates, Samuel Rubin Foundation, United States Institute of Peace, World Bank, and individual contributors.

Susan Higman
Managing Editor
Associate Director
Student Pugwash USA
September 1996

Table of Contents

Foreword — i

Preface
A Message From Professor Joseph Rotblat, — ii
   1995 Nobel Peace Laureate
A Promise for the Future — iv

Section 1: Get a Job! — 1
Preparing Your Application Materials — 3
Cover Letters — 4
Resumes — 5
Often-Requested Information — 6
Preparing for Your Interviews — 7
The Interview — 8
General Resources for the Job Hunt — 10

Section 2: Where Should I Work? — 13
Organizational Profiles — 15

Section 3: Where Else Can I Look? — 245
Opportunities with the Government — 247
International Opportunities — 251
Researching Corporate Responsibility — 254

Section 4: More About Student Pugwash USA — 257

Section 5: Indices — 261
Organizations at a Glance — 262
Geographical Listing — 288

Foreword

Welcome to the hunt for the perfect job or internship! At the crossroads of science, technology, and society are a myriad of dynamic organizations working to make the world safe, sustainable, and peaceful. They bring together people of different talents, ideologies, educational experiences, and cultural backgrounds. As we approach the 21st century, organizations focusing on science and technology are not just for those with PhDs in physics or chemistry. They are also for people with diverse backgrounds in political science, accounting, graphic design, and editing. The search you are undertaking may lead you to find your life's work or a new career path. *Jobs You Can Live With* will help you research your options and acquaint you with organizations for which you might consider working.

Student Pugwash USA produced this fifth edition of our directory to fill a niche—a guide to finding a job at the juncture of science, technology, and society. This book features organizations that are working to explore the ethical, social, and global implications of science and technology. *Jobs You Can Live With* offers an introduction to organizations that work in the fields of energy, environment, food and agriculture, global security, health and biomedical research, and information technologies. Many of these organizations offer staff and/or internship positions.

In addition, this book seeks to inspire you with the words of the 1995 Nobel Peace Prize recipient, Professor Joseph Rotblat. Professor Rotblat shared the Peace Prize with the Pugwash Conferences on Science and World Affairs, which he founded in 1957. Through his decision to leave the Manhattan Project before the completion of the atomic bomb and his commitment to working for peace, Professor Rotblat demonstrated that we can all make a difference.

The transition from college to full-time employment is a challenge. Work can be both exhausting and rewarding—an environment where new skills are acquired as much out of necessity as curiosity and where working on exciting issues and progressing toward a goal are part of the remuneration.

Making a better world requires people to work together to bring about positive change. Part of that dream is for people to enjoy their work—it's hard to imagine peace without happiness. We wish you well on your journey to find, not only the perfect job, but a career in which you are challenged and encouraged. We hope you succeed beyond your wildest expectations!

Preface

A Message from Professor Joseph Rotblat,
1995 Nobel Peace Prize Recipient

There was a time when scientists lived in ivory towers, pretending that their work had nothing to do with societal problems. The aim of science—they claimed—was to understand the laws of nature, and since these are immutable and unaffected by human reactions and emotions, these reactions and emotions are not the concern of scientists.

This was a fallacy and illusion even before the recognition of the "social sciences" as a legitimate branch of scientific pursuit. But the crumbling walls of the ivory towers were finally demolished by the blast wave of the Hiroshima bomb. Science not only affects human relations; the whole fate of humankind is affected by science.

Half a century into the nuclear age we should remind ourselves of the chief characteristic of that age: for the first time in the history of civilization it had become possible for Man to destroy the whole of civilization in one go. On several occasions we have come very close to the edge of the abyss. It was more by good luck than by good management that we have avoided catastrophe. With the end of the Cold War, the likelihood of a nuclear holocaust has been greatly reduced, but it has not been reduced to zero. The threat of a nuclear confrontation will be real as long as nuclear weapons exist.

The nuclear issue is not the only mortal peril we face; our civilization may be heading to extinction by yet another consequence of science and technology; the steady poisoning of the environment. Whether in a bang or with a whimper, civilization will be doomed unless a conscious effort is made to save it. It is incumbent on scientists to be in the forefront of that effort.

In this respect, the positive achievements of science and technology are of great importance. The tremendous advances in communication and transportation have provided the means to create a true world community in which all members are strongly interdependent. We have the means to establish close cultural bonds and get to know each other better thanks to the ease of travel and direct access to information. All this should help to overcome chauvinism and xenophobia—the fomenters of strife and war. We have now the tools to foster the feeling of belonging to humankind, in the spirit of the Russell-Einstein Manifesto when it said: *"We are speaking . . . not as members of this or that nation, continent, or creed, but as human beings, members of the species Man, whose continued existence is in doubt."*

Scientists have a substantial role to play in this, because the feeling of belonging to a world community is already well developed in them by the fundamental characteristic of science; universality. The second characteristic, openness, is also highly relevant, doing away with secrecy is an essential part of the efforts to abolish war. Scientists can take a lead in these efforts, as well as serving as a model for the rest of the community.

The Pugwash Movement is an expression of the need of scientists to fulfill their social and moral responsibilities, to take an active part in the creation of a peaceful world. This applies both to senior scientists already established in their professions, and to the young scientists on the threshold of their careers. In choosing a field of work, ethical issues must be given due consideration. For some young scientists this may be in the form of a pledge not to use their scientific training for any purpose intended to harm human beings—a kind of Hippocratic Oath. With or without a formal pledge, it is incumbent on all young scientists to ponder on the implications of their work and ensure responsible use of science and technology.

For those in the social sciences, the nature of their work usually keeps them aware of the societal implications. This is not generally so for the natural scientists. The ever growing specialization of science demands concentration on very narrow sectors of knowledge, and this often leads to a loss of wider horizons. To remedy this, scientists should make a deliberate decision to devote a small fraction of their time, say five to ten per cent, to the fulfillment of their obligations to the community. An involvement in Student/Young Pugwash activities, by attending meetings or debating issues via computer network, is a satisfactory way to meet these obligations.

Although the ideological struggle that divided the world for most of this century is largely over, there are still many causes of strife; there are still many bloody wars; far too many weapons in the arsenals; an ever-growing assault on the natural environment; increasing disparity in standards of living between nations and between people. We scientists, young and old, must nurture a vision of a better world in the next century, a world without war, a society based on care and equity, a community that will protect the environment. And we should make it our task to turn this vision into reality.

Joseph Rotblat
President, Pugwash Conferences on Science and World Affairs

The Pugwash movement began on July 9, 1955, when eminent scientists and scholars, led by Albert Einstein, Bertrand Russell, and Joseph Rotblat called on scientists and the public to abandon the notion of international violence as a political tool. Since 1957, leading scientists and scholars have met annually at the Pugwash Conferences on Science and World Affairs to discuss critical global issues. On December 10, 1995, the Norwegian Nobel Committee awarded the 1995 Nobel Peace Prize to the Pugwash Conferences on Science and World Affairs and their president and founder, Professor Joseph Rotblat, for their efforts to "diminish the part played by nuclear arms in international politics and in the longer run to eliminate such arms." Student Pugwash USA was founded in 1979 and is one of approximately 20 national Student and Young Pugwash groups around the world.

A Promise for the Future

In celebration of the Nobel Peace Prize, Student Pugwash USA has taken up Professor Rotblat's suggestion and developed a pledge that advocates the responsible use of science and technology. This exciting initiative to get commitments from students and young professionals is striking a positive chord as people from around the world are adding their names to those who have signed the pledge:

I promise to work for a better world, where science and technology are used in socially responsible ways. I will not use my education for any purpose intended to harm human beings or the environment. Throughout my career, I will consider the ethical implications of my work before I take action. While the demands placed upon me may be great, I sign this declaration because I recognize that individual responsibility is the first step on the path to peace.

As you begin your professional career you might consider taking this pledge. This campaign is not limited to students or scientists. We can all commit to making sure science and technology are used responsibly. After all, science and technology touch everyone's life! Each and every signature brings us one step closer to a more ethical and peaceful world.

For more information, call our office at 1-800-WOW-A-PUG, send us e-mail at pledge@spusa.org, or visit our Web page at http://www.spusa.org/pugwash/pledge.html.

Section 1
GET A JOB!

Getting a job is not easy. For those who have just left the halls of academe and are entering the job market, this is a time of new experiences and adventures as you try to take the world by storm. Job hunting is stressful, especially if you are trying to find a job that you can believe in. For those looking for an internship, this is a time to explore different avenues as you try to decide what to do with your life.

To assist you in landing a job, this section of *Jobs You Can Live With* contains helpful hints on:
- Preparing your application materials
- Preparing for an interview
- Where to look for additional information

Preparing Your Application Materials

Generally, the materials you submit to an organization or business will give potential employers their first impression of you. Each piece should be typed and free of spelling and grammatical errors. Although it may seem obvious, proofreading and spell checking are extremely important, as they indicate your attention to detail and may determine whether or not you are called for an interview.

Any materials you submit should arrive before the deadlines. If the deadline has passed, you may want to call to see if they are still accepting applications. To mail the materials, it is best to use a large envelope to ensure they arrive in pristine condition. If you would like to deliver your information to the organization in person, call and verify that the contact person (or someone else) will be available. A visit to the office presents an opportunity to ask for a complete description of the job and see the work environment. However, a practical drawback to visiting the office (especially unannounced) is that the timing may be inconvenient, meaning that no one can meet with you.

The following list identifies the most commonly requested materials. A more detailed description of each element is provided on the following pages.

Cover Letters—introduce you to the organization and should always be included.

Resumes—summarize your academic and professional experience.

Applications—are specific forms some places ask you to fill out.

Writing Samples—give insight into your abilities.

Transcripts—are either informal or official copies of your coursework that may be requested.

References and Recommendations—can be provided by people who will endorse and verify your relevant abilities.

Cover Letters

Your cover letter will give potential employers an idea of why you want to work for them. Each cover letter should be written specifically for the organization and position sought. It should serve as an introduction of who you are and indicate why you are interested in the organization and in the specific job for which you are applying. Your cover letter and resume should complement each other, with the letter expanding upon the experiences and skills highlighted in the resume.

The tone of the cover letter has a significant impact on its reader. Although the cover letter is a marketing tool, it is important not to come on too strong with the hard-sell approach. A well-crafted cover letter that highlights your most valuable skills and displays an understated confidence will often encourage the employer to pay more attention to your resume. On the other hand, a poorly written or sloppily presented letter can dissuade a potential employer from seriously considering you.

Here are some basic points to remember in writing a strong cover letter:
- **Proofread, spell-check, and proofread again.** Few things make a worse impression than cover letter typos. Set the letter aside for a day before doing the final read or pass it on to a friend, colleague, or advisor who will give you constructive feedback.
- **Get the organization's contact information right.** Call before sending your materials to verify the contact person's name and the organization's address. Always use a gender-neutral salutation if you do not have the contact person's name.
- **Keep the letter brief.** Three or four concise paragraphs should be plenty. Too much information all at once is not necessarily a good thing.
- **Explain who you are and why you want to work with the organization.** Focus on the skills or experiences that pertain to the position, including academic, professional, or volunteer activities. It is also acceptable to include your enthusiasm for the organization, the job, and the opportunity for personal and professional growth. A career counselor can help you outline pertinent experiences or skills.
- **Focus on your three most impressive relevant skills and experiences.** Show them how your experience and creative ideas can benefit their organization.
- **Indicate your general period of availability.** This is helpful for jobs that are of unspecified duration.
- **If you are not local and plan to visit the organization's area, tell them.** You may be able to arrange an interview.
- **Mention that you will call the organization by a specific date.** And then call.
- **Always provide your correct address and telephone number.** Do not make a potential employer track you down. Indicate when and where you can be reached.
- **Don't forget to sign the letter.** It happens and always looks bad.

Resumes

Your resume summarizes relevant skills and experiences and provides a foundation for comparison with other job candidates. Both the content and visual appearance of your resume communicate important messages to the reader, so strive to make your resume as informative and well-organized as possible.

Your resume should be typed, never handwritten. Resumes should be as reader-friendly as possible. To facilitate its reading, selective use of highlighting, italics, fonts, and other easy-to-read markers can be helpful, but be careful not to go overboard. Professionalism is important. Sloppiness or typographical errors may result in your resume being relegated to the bottom of the pile.

There are many styles of resumes. Some people strongly adhere to the single page rule while others prefer an enhanced description of your experiences, such as a curriculum vitae (CV). It may be helpful to look at sample resumes in order to get an idea of the best way to present yourself, or to work with a career counselor to develop your resume.

Here are some suggestions about developing an effective resume:
- **Proofread, spell-check, and proofread again.** You have heard it before, but it's important. As with your cover letter, have someone proofread your resume for clarity and grammar.
- Make sure that your address and telephone number are correct.
- **If you use an "Objective" statement (optional), tailor it specifically to each position.** Objective statements should only be used for specific jobs and generic objectives should be avoided. "Job objectives" indicate your goals to potential employers. A "summary" tells what you can do for them. A note of caution: an employer's objectives or goals may be slightly different than yours.
- **If your GPA is above 3.0, include it.** If it is below 3.0, do not include it.
- **List academic and professional experiences in chronological order.** The key is to make the resume flow. After college, omit high school-related experiences unless they are very compelling.
- **List relevant volunteer or extracurricular experiences.** This is useful to show a well-rounded quality and may bring out abilities that are not apparent from your academic or professional experiences.
- **Omit personal information.** Family details and hobbies are not considered professionally relevant.
- **Keep description of experiences brief.** Since job titles are not universal, describe your experiences and try to make them sound as germane as possible. On a resume, five lines per position is an acceptable maximum length; on a CV, descriptions should be less than ten lines per position.
- **Use action words.** You want to convey confidence, growth, and experience. But be careful, don't oversell yourself—remember, you have to live up to your resume if you get the job!

Often-Requested Information

Applications: While application forms vary from organization to organization, use the opportunity to emphasize the ways in which your skills and experiences can meet the needs of the organization. Think about why they are asking each question and what kind of information they are looking for.

Writing Samples: Many organizations request writing samples to evaluate your writing style, editing capabilities, or researching skills. The specific content is generally less important than the quality of work. Most organizations prefer short essays, academic papers, published articles, or professional analyses that deal with issues relevant to their specific programs and issues. One to four pages is usually sufficient.

Transcripts: Some organizations require that you provide either informal or official transcripts from all post-secondary schools. In general, this is to confirm a level of academic competence in the area of the job's focus. For example, a student may have a GPA of 3.5 overall, but only 2.8 in a specific field of study related to the job.

References and Recommendations: Many organizations request the names of individuals who can evaluate your professional or academic performance. Naturally, you will want to provide the names of people who will comment favorably on your abilities. Before you list people as references, be sure to inform them so that they will not be surprised if prospective employers call. Call your references in advance and provide them with a general description of the organization and the specific details concerning the position for which you are applying. Make them aware of any skills or experiences that you would like them to emphasize. They can then provide a strong, supportive recommendation, commenting on your knowledge of relevant issues, your abilities as a student or worker, and your willingness to learn and assume responsibility. As with your own contact information, it is important to provide accurate phone numbers and addresses for each reference, and if possible, the best times to reach them.

Some organizations prefer to review written recommendations instead of calling references. If this is the case, contact your references and ask them to write letters of recommendation on your behalf. As above, provide references with information about the organization, position, and points to highlight. Supply your references with a stamped, addressed envelope and stress to them the importance of meeting the organization's deadline for recommendations. You should also provide the organization with the name and address of each person submitting a letter of reference on your behalf. It is sometimes helpful to have a general letter of recommendation to submit with your resume and cover letter, but this is not essential.

Preparing for Your Interviews

Interviews provide the opportunity for you and the interested organization to learn more about each other. While most obviously a chance for employers to investigate your experiences and skills, the interview is also an excellent opportunity for you to find out more about the organization in terms of its activities and its working environment. Interviews take one of two forms. Employment interviews are conducted to fill a vacancy in the organization. Informational interviews are conducted to give the applicant a better understanding of the organization.

In order to get the most out of an interview, you will need to do a little homework. Research the organization. Develop some thoughtful questions you can ask the interviewer about the organization's mission, staffing structure, and current job openings. You should also think through the types of the questions that you may be asked.

Employment Interviews: Interviews for employment in an organization may be quite stressful. Of course, the basic rules are to dress in an appropriately professional manner and to be on time or about 10 to 15 minutes early. In addition, be prepared to be interviewed by more than one person, either concurrently or consecutively.

Try to give full answers to questions, i.e., relate your experiences to the job or cite a class you attended where you learned helpful hints. Focus on what you know or what you are interested in. Put a positive spin on your abilities. Remember all the dimensions of the job and try to address them in your answers. Emphasize your strengths, such as organizing ability, working with people, research experience, writing ability, creativity, or some of the "extra" items on your resume (public relations, media, or computer experience). Show the organization why you stand out from the crowd. It is also helpful to bring copies of materials submitted (resume, cover letter, etc.) and other information the interviewer may be interested in seeing.

Informational Interviews: While similar in many ways to employment interviews, informational interviews focus on the exchange of information rather than on evaluation. This process allows you to learn more about a particular organization or field and gain experience in interviewing without the stress of being evaluated. It is important to remember that informational interviews are conducted almost exclusively for the benefit of the interviewee. As with employment interviews, be prepared with the questions you would like to ask the interviewer. Doing your homework will help the meeting flow smoothly and will impress upon the interviewer that you consider her or him a valuable resource person and appreciate their time.

The Interview

During an interview, you will be asked a series of questions so the interviewer can get to know you and have an idea of your abilities. In both employment and informational interviews, there are some standard types of questions that you should be prepared to answer:

• **Your background, both academic and extra-curricular**. This will tell the interviewer the ways in which your academic, volunteer, and professional skills and experiences would benefit the organization. Such questions may address your favorite (or least favorite) job or project, the ways in which you describe yourself, and your strengths and weaknesses.
• **Your interest in the organization and the position.** The interviewer will want to know why he or she should hire you. These questions may focus on your reasons for wanting to work there, the elements of the job description that most appeal to you, and what aspects of the job would challenge you.
• **Your career goals**. The interviewer will want to get a sense of how you see your future. Don't worry if your goals are vague, just have some ideas. Be prepared to show how your goals relate to the position in question. Likely questions may touch on your aspirations for the future (graduate school, start your own business, teach, do research, etc.), your vision of yourself in five or ten years, and the ways in which you expect to benefit from the job and the organization.
• **Traits that you possess.** The interviewer may ask some questions about your work style to get an idea of how you would fit in with the rest of the staff. Questions may address the ways in which you deal with stress, your time management and prioritizing skills, and your ability to manage the details without losing sight of the overall picture.

You may want to ask some questions, too:

• **About the position.** Among the issues you may want to address—the duration of the position, the average length of stay for employees, potential opportunities to stay on longer or advance within the organization, salary and benefits (such as health care, educational opportunities, and investment possibilities), information about the supervisors, and when you might expect to hear from them about their decision.
• **About working in the organization.** Questions to help you get a sense of the office atmosphere include descriptions of an average day, the educational and experiential requirements for other positions, the most enjoyable part of the interviewer's work, the reasons why the interviewer likes working there or one thing they would change about the organization.
• **About the organization**. In order to better understand the organization, you may ask about the primary focus of their work, to see publications (newsletter, issue brief, or annual report), the number and target group of their membership, and a list of their supporters and board of directors.

There are also questions you cannot be asked in an interview. For example, you cannot be asked direct questions about your religion, balancing career and family, or your health. However, indirect questions are allowed when posed to all applicants. For example, "Are you Jewish and therefore unable to work on Saturdays?" is not allowed, however "This job may require working on weekends. Is that acceptable?" is appropriate when posed to all candidates. If you are interested in finding out more about the legal aspects of interviews, there are resources that deal with these issues in most bookstores.

Following the interview, you will want to send a letter of appreciation and call to see if there is any additional information required. However, keep in mind that the hiring process is time-consuming. Be efficient and brief. Before you call, make a list of questions you want to ask or points you would like clarified.

General Resources for the Job Hunt

Here are some additional resources that should help with your general job and internship search and that feature opportunities in a number of different areas:

100 Best Careers for the 21st Century (1996).

100 Best Careers for the Year 2000 by Shelly Field (1992).
Whole Work Catalog (800) 634-9024.

The 100 Best Companies to Work for in America by Moskowitz, Levering, and Katz (1993). Whole Work Catalog (800) 634-9024.

The Berkeley Guide to Employment for New College Graduates by James I. Briggs (1984). Ten Speed Press (510) 845-8414.

The Career Discovery Project by Dr. Struman (1993).
Whole Work Catalog (800) 634-9024.

The Career Finder: Pathways to Over 1500 Entry-Level Jobs by Schwartz (1990).
Whole Work Catalog (800) 634-9024.

Good Works: A Guide to Careers in Social Change.

Green At Work by Susan Cohn (1995). Island Press (800) 828-1302.

How to Get the Job You Want: For Unskilled, Skilled, and Professional (1994).
Zinks Publisher.

How to Locate Jobs and Land Internships by Albert French (1993).
Whole Work Catalog (800) 634-9024.

Making a Living While Making A Difference, A Guide to Creating Careers with a Conscience by Melissa Everett (1995). Bantam Books.

The Minority Career Book by Dr. Rivera (1991).
Whole Work Catalog (800) 634-9024.

The New Complete Guide to Environmental Careers (1993).
Island Press (800) 828-1302.

Non-Profits' Job Finder by Daniel Lauber (1995).
Career Communications (800) 346-1848.

Professional's Private Sector Job Finder by Daniel Lauber (1995).
Career Communications (800) 346-1848.

Put Your Degree to Work by Marcia R. Fox (1988).
W.W. Norton, New York (212) 354-5500.

Researching Your Way to a Good Job by Crowther (1993).
Whole Work Catalog (800) 634-9024.

The Smart Woman's Guide to Career Success by Janet Hauter (1993).
Whole Work Catalog (800) 634-9024.

The Twentysomething Guide to Creative Self-Employment by Jeff Porten (1996).
Prima Publishing (916) 632-4400.

What Color is Your Parachute by Richard Nelson (1993).
Ten Speed Press (510) 845-8414.

Who's Hiring Who? How to Find That Job Fast by Richard Lathrop (1989).
Ten Speed Press (510) 845-8414.

Job Sites on the Web

There are several places on-line that offer job hunt advice and options. A few such sites are listed below.

The Monster Board - http://www2.monster.com/home.html
- career center
- job listings
- links to other sites

America's Job Bank - http://www.ajb.dni.us/
- state employment services

The Internet's On-Line Career Center - http://www.occ.com/occ/

Fed World Federal Job Search - http://www.fedworld.gov/jobs/jobsearch.html

Federal Jobs Digest - http://www.jobsfed.com/

Adams Job Bank Online - http://www.adamsjobbank.com/
- resources

Career Mosaic - http://www.careermosaic.com/cm
- resources
- links to other sites
- international sites

Career Paths. com - http://www.careerpath.com
- searches newspapers

Help Wanted USA - http://lccweb.com/employ.html

Job Web - http://www.jobweb.org/
- resources
- links to other sites
- international sites

Virtual Job Fair - http://www.vjf.com/
- resume center & other resources

Job Bank USA - http://www.jobbank usa.com
- links to international job bank
- links to other sites

Section 2
WHERE SHOULD I WORK?

This section will introduce you to organizations that work at the crossroads of science, technology, and society. The featured organizations all work in at least one of the following broad categories, represented by a series of graphic symbols for easier recognition.

We encourage you to read through all the profiles in order to get a better idea of your range of options. If you are interested in one particular area, there are indices at the back of the book to which you can refer.

We have listed these organizations to give you a starting point on your career search. It is not a comprehensive list by any means. The information listed was provided to us by each organization. We cannot guarantee its veracity. Do your homework and good luck!

Please note: The name of a contact person is not given because, based on our information, those names become outdated more rapidly than the other organizational information. Please call the organization for the name and title of the person to whom materials should be sent. If an organization prefers not to receive calls, send materials to "Internship Coordinator" or "Director of Personnel."

20/20 Vision

Address: 1828 Jefferson Place, NW
City: Washington   **State:** DC   **Zip:** 20036

Telephone: (202) 833-2020   **Fax:** (202) 833-5307
E-mail: vision@igc.apc.org
URL: http://www.2020vision.org/

Other locations: None listed

Founded: 1986

Mission:
20/20 Vision is devoted to the revitalization of democracy through citizen action, focusing primarily on protecting the environment and working for peace through nuclear arms control, decreased military spending, and reduced arms sales.

Programs:
A monthly legislative update (10 to 15 pages) describing hot issues in Congress is distributed to several hundred activists nationwide. Action alert postcards on related issues and occasional fact sheets are produced for grassroots activists.

Atmosphere: Casual, hectic, occasional fun activities

Volunteers: Over 100

STAFF POSITIONS
Full-time staff: 6   **Part-time staff:** 2
Staff positions available: 1 to 2 per year
Staff salary: Fellowship, $10,000; others, low- to mid-twenties

INTERNSHIPS
How Many: 3   **Duration:** 4 months
Remuneration: Unpaid
Duties:
The work is a mix of administrative and substantive. Interns can choose to work on legislative and policy issues, promotions, membership, or development.

Qualifications:
An interest in the issues and the ability to work with limited supervision.

How to apply: Cover letter, resume, writing sample, and references

**JOBS
YOU CAN
LIVE WITH
1996**

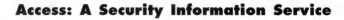

Access: A Security Information Service

Address: 1511 K Street, NW, Suite 643
City: Washington **State:** DC **Zip:** 20005-1401

Telephone: (202) 783-6050 **Fax:** (202) 783-4767
E-mail: access@4access.org
URL: http://www.4access.org

Other locations: None listed

Founded: 1985

Mission:
Access is a nonprofit, nonadvocacy information service on international security, peace, and world affairs. In an effort to inform the public of important international developments, Access publishes impartial materials with clear summaries of current issues and overviews of diverse perspectives in the foreign policy debate. Access endorses no position but gives reference to sources that represent various views in order to expand and inform public debate.

Programs:
Access maintains various databases on international affairs organizations, ethnic conflicts, and conflict resolution. The international affairs database contains approximately 2500 organizations around the world working on these issues. Access operates a resource and referral service. There are also numerous related publications produced by Access.

Atmosphere: Semiformal, hectic at times

Volunteers: 0

STAFF POSITIONS
Full-time staff: 5 **Part-time staff:** 1
Staff positions available: Usually none
Staff salary: Varies

INTERNSHIPS
How Many: Varies **Duration:** Minimum of 8 weeks
Remuneration: $50 weekly, academic credit
Duties:
Research and write for issue papers, assist with production of issue guides and directories, and interns are assigned an administrative task to be completed throughout the internship.

Qualifications:
Interns need to be at least a junior in college with relevant fields of interest.

How to apply: Cover letter, resume, and writing sample

ACCESS: Networking in the Public Interest

Address: PO Box 65881
City: Washington **State:** DC **Zip:** 20035-5881

Telephone: (202) 785-4233 **Fax:** (202) 785-4212
E-mail: Not listed
URL: Not listed

Other locations: None listed

Founded: 1987

Mission:
ACCESS: Networking in the Public Interest is a nonprofit organization making it easier for people to become involved in addressing the public challenges they care about most. Since its founding, ACCESS has served as the nation's only comprehensive clearinghouse of information regarding employment opportunities in the nonprofit sector.

Programs:
Community Jobs—a national employment newspaper for the nonprofit sector. Special editions are available for the New York-New Jersey area and Washington, DC.
Non-Profit Career Development and Resume Review—available by phone nationwide.
Non-Profit Organization Search.
College Package of Community Jobs.
Opportunities in Public Interest Law.
The National Service Guide—how to find opportunities to make a difference.

Atmosphere: Always casual, often hectic

Volunteers: 0

STAFF POSITIONS
Full-time staff: 3 **Part-time staff:** 1
Staff positions available: 1 per year

Staff salary: Varies, depending on experience and education

INTERNSHIPS
How Many: Varies **Duration:** As needed
Remuneration: Based on duties, experience
Duties:
Assist in office, desktop publishing, and general outreach.

Qualifications:
Initiative, knowledge, concern for social justice and service, office experience, and computer literacy.

How to apply: Phone call

ACCION International

Address: 120 Beacon Street
City: Somerville   **State:** MA   **Zip:** 02143

Telephone: (617) 492-4930   **Fax:** (617) 876-9509
E-mail: 72420.2015@COMPUSERVE.COM
URL: Not listed

Other locations: Washington, DC and Tucson, AZ

Founded: 1961

Mission:
ACCION International is dedicated to reducing poverty and unemployment throughout the Americas by providing basic business training and small, short-term loans to owners of the smallest-scale businesses, or "microenterprises," through a network of affiliated organizations. By providing these business tools, ACCION works to increase family income and create new jobs in low-income communities from Santiago to New York.

Programs:
ACCION works with local affiliated institutions in 13 Latin American countries and 6 US cities. ACCION provides microenterprise support in the form of training and financial services, in order to increase the self-reliance and well-being of small-scale entrepreneurs. ACCION has a series of publications regarding the learning and best practices in the field of microenterprise assistance. Catalogs containing outlines may be purchased. ACCION also publishes an annual report and quarterly bulletins.

Atmosphere: Hectic and informal

Volunteers: 5

STAFF POSITIONS
Full-time staff: 35   **Part-time staff:** 5
Staff positions available: 1 per year
Staff salary: $15,000 to $50,000 per year

INTERNSHIPS
How Many: Varies   **Duration:** Varies
Remuneration: Negotiable
Duties:
Depends on project and intern interest.

Qualifications:
At least a bachelor's degree.

How to apply: Cover letter and resume

Advocacy Institute

Address: 1707 L Street, NW, Suite 400
City: Washington   **State:** DC   **Zip:** 20036

Telephone: (202) 659-8475   **Fax:** (202) 659-8484
E-mail: Not listed
URL: Not listed

Other locations: None listed

Founded: 1985

Mission:
The Advocacy Institute is devoted to strengthening the capacity of social and economic justice advocated to influence social and public policy.

Programs:
1. Issue laboratories to study and experiment with advocacy tools, and work with grassroots advocates to strengthen the movement's infrastructure; focus on tobacco control, sustainable transportation, and media advocacy in response to libertarian rhetoric of the Right.
2. Advocacy and capacity building training programs work with advocates from the US and abroad to strengthen organizations' and individuals' abilities to affect public policy and social change.
3. Computer networking to explore and determine the usefulness of emerging technologies and share these lessons with advocates for progressive social change.

Atmosphere: Casual and hectic

Volunteers: 0

STAFF POSITIONS
Full-time staff: 30   **Part-time staff:** 3
Staff positions available: Several every year
Staff salary: $22,000 per year to start

INTERNSHIPS
How Many: Varies   **Duration:** 10 to 12 weeks
Remuneration: $1,000 per month
Duties:
Substantive programmatic work and support.

Qualifications:
College student or graduate and a commitment to public interest.

How to apply: Call to request application packet

African Scientific Institute

Address: PO Box 12161
City: Oakland **State:** CA **Zip:** 94604

Telephone: (510) 653-7027 **Fax:** (510) 547-0387
E-mail: Not listed
URL: Not listed

Other locations: None listed

Founded: 1967

Mission:
The African Scientific Institute is a nonprofit, tax-exempt corporation representing a network of scientists, engineers, technologists, and health professionals, as well as young people aspiring to enter the world of science and technology.

Programs:
The African Scientific Institute programs are: publishing related literature; sponsoring conferences, seminars, workshops, researching and consulting services, and lectures before various types of audiences; making radio and television appearances; counseling and providing special support services for scientifically motivated youth; coordinating community involvement projects, such as Achievement Award Banquets and co-operative projects with private and governmental sectors; and employment matching programs.

Atmosphere: Casual

Volunteers: 200

STAFF POSITIONS
Full-time staff: 0 **Part-time staff:** 0
Staff positions available: None
Staff salary: Not listed

INTERNSHIPS
How Many: 20 **Duration:** 9 months
Remuneration: Commission, based on project
Duties:
Contact notable scientists, develop funding methodologies and proposals, write press releases and news stories.

Qualifications:
At least two years of college course work in science and technology and a sensitivity towards minorities.

How to apply: Cover letter and resume

Alaska Green Goods

Address: 3535 College Road
City: Fairbanks   **State:** AK   **Zip:** 99709

Telephone: (907) 452-4426   **Fax:** (907) 452-4136
E-mail: Not listed
URL: Not listed

Other locations: None listed

Founded: 1992

Mission:
Alaska Green Goods is a retail store featuring environmentally sound products, including efficient lighting, organic cotton clothing, hemp products, personal care items, and gifts.

Programs:
The store promotes sustainable consumer habits by offering alternative products and education.

Atmosphere: Casual

Volunteers: Varies

STAFF POSITIONS
Full-time staff: 1   **Part-time staff:** 1

Staff positions available: 1 to 2 per year

Staff salary: Depends on experience

INTERNSHIPS
How Many: Varies   **Duration:** Varies
Remuneration: Depends on experience
Duties:
Interns work to create special programs in education and marketing.

Qualifications:
An interest in business and the environment, professional appearance, flexible, and an open mind.

How to apply: Cover letter and resume

JOBS YOU CAN LIVE WITH 1996

Alice Hamilton Occupational Health Center

Address: 408 7th Street, SE
City: Washington **State:** DC **Zip:** 20003

Telephone: (202) 543-0005 **Fax:** (202) 543-1327
E-mail: Not listed
URL: Not listed

Other locations: Landover, MD

Founded: 1986

Mission:
The Alice Hamilton Occupational Health Center is dedicated to providing training, technical assistance, research and legislative advocacy services in the area of occupational safety and health. The Center operates primarily in the DC metropolitan area.

Programs:
Major programs involve asbestos, lead, and hazardous materials training. The Center also distributes the EPA Asbestos Manual and the EPA Lead Manual to workers and instructors.

Atmosphere: Causal

Volunteers: 3 to 4

STAFF POSITIONS
Full-time staff: 7 **Part-time staff:** 2
Staff positions available: Not listed
Staff salary: Not listed

INTERNSHIPS
How Many: 0 **Duration:** Not listed
Remuneration: Not listed
Duties:
Not listed
Qualifications:
Not listed
How to apply: Not listed

Alternative Energy Resources Organization (AERO)

Address: 25 S. Ewing, Suite 214
City: Helena   **State:** MT   **Zip:** 59601

Telephone: (406) 443-7272   **Fax:** (406) 442-9120
E-mail: aero@desktop.org
URL: Not listed

Other locations: None listed

Founded: 1974

Mission:
AERO is a grassroots organization working to improve the quality of peoples' lives, communities and environment. The Montana-based, nonprofit organization serves the needs of people working to promote renewable energy and conservation, sustainable agriculture, and greater community self-reliance.

Programs:
Sustainable Agriculture/Communities—aids Northern Plains and Rockies farmers and ranchers in finding ways to sustain farm productivity and maintain rural communities and natural resources.
Energy Conservation and Renewable Resources—works with policymakers and communities to develop valid options; provides energy-saving information to homeowners and businesses.
Family Farm Transition—helps beginning farmers in Montana get started and helps to ease the transition of those retiring.
Montana Transportation Project—organizes groups and activists, brings peoples' voices into shaping transportation policy.

Atmosphere: Casual, relaxed, good morale

Volunteers: Varies

STAFF POSITIONS
Full-time staff: 5   **Part-time staff:** 3
Staff positions available: Varies
Staff salary: $16,000 to $24,000 per year

INTERNSHIPS
How Many: 1 to 2   **Duration:** Varies, one semester
Remuneration: Currently unpaid
Duties:
Program assistance and development.

Qualifications:
Competence in work environment and office functions; knowledge of sustainable agriculture, renewable energy, community development; good academic standing; teamwork and reliability.

How to apply: Cover letter, resume, and writing sample

**JOBS
YOU CAN
LIVE WITH
1996**

American Forests

Address: 1516 P Street, NW
City: Washington **State:** DC **Zip:** 20005

Telephone: (202) 667-3300 **Fax:** (202) 667-7751
E-mail: Not listed
URL: Not listed

Other locations: None listed

Founded: 1875

Mission:
American Forests works to protect and conserve all the benefits of trees and forests for present and future generations.

Programs:
American Forests magazine, Global ReLeaf (tree planting, ecosystem restoration programs), Forest Policy Center, urban forestry, Citizen Forestry Support Center.

Atmosphere: Casual and focused

Volunteers: 5

STAFF POSITIONS
Full-time staff: 40 **Part-time staff:** 0
Staff positions available: 1 to 3 per year
Staff salary: $20,000 to $30,000 per year

INTERNSHIPS
How Many: 4 to 6 **Duration:** Varies
Remuneration: Possible stipend
Duties:
Administration support, direct program work, and hands-on environmental work.
Qualifications:
Education, writing ability, and environmental ethics.

How to apply: Cover letter and resume

American Hydrogen Association

Address: 216 S. Clark Drive, #103
City: Tempe   **State:** AZ   **Zip:** 85281

Telephone: (602) 921-0433   **Fax:** (602) 967-6601
E-mail: aha@getnet.com
URL: http://www.getnet.com/charity/aha

Other locations: None listed

Founded: 1989

Mission:
The Association works to promote nonpolluting renewable resources, especially hydrogen fuel made from solar energy, and to encourage economic prosperity without pollution. This mission will be achieved through education and scientific activities, such as publications, research, demonstrations, public presentations, and sales of equipment and services that support renewable resources.

Programs:
Currently, the American Hydrogen Association is demonstrating the prototypes of the hydrogen economy and creating a market demand. Business people are adding hydrogen to their business plans and outlining projects that, one farm, one home, and one factory at a time, can become energy independent. They also publish *Hydrogen Today*, the official newsletter of the Hydrogen Association.

Atmosphere: Casual

Volunteers: All are volunteers

STAFF POSITIONS
Full-time staff: Varies   **Part-time staff:** Varies
Staff positions available: None
Staff salary: All volunteer

INTERNSHIPS
How Many: Varies   **Duration:** Varies
Remuneration: Unpaid
Duties:
Varies. The Association is translating materials into Japanese and needs interns in advertising, public relations, newsletter production, conversion/mechanical, and other areas. Duties depend on the talents of the individual.

Qualifications:
Motivation and a knowledge of how to help the organization.

How to apply: Cover letter and resume; phone calls welcome

Antarctica Project

Address: 424 C Street, NE
City: Washington **State:** DC **Zip:** 20002

Telephone: (202) 544-0236 **Fax:** (202) 544-8483
E-mail: antarctica@igc.apc.org
URL: Not listed

Other locations: None listed

Founded: 1982

Mission:
The Project works to preserve Antarctica by monitoring activities to ensure minimal environmental impact and consulting with key users, including scientists, tourists, and governments. The Project is secretariat to the Antarctic and Southern Ocean Coalition (230 conservation groups in 50 nations), conducts legal and policy research and analysis, produces educational materials, and focuses on the scientific community and globally significant research.

Programs:
Projects vary, but there is a quarterly newsletter provided to all members.

Atmosphere: Professional, casual attire, somewhat hectic

Volunteers: Unlimited

STAFF POSITIONS
Full-time staff: 2 **Part-time staff:** 0
Staff positions available: Usually none
Staff salary: Varies

INTERNSHIPS
How Many: 3 per semester **Duration:** 3 to 4 months
Remuneration: Unpaid
Duties:
Depends on background and interest.

Qualifications:
Motivated, independent, and capable.

How to apply: Cover letter, resume, and writing sample

Appalachian Mountain Club

Address: 5 Joy Street
City: Boston   **State:** MA   **Zip:** 02108

Telephone: (617) 523-0636   **Fax:** (617) 523-0722
E-mail: Not listed
URL: Not listed

Other locations: Gorham, NH

Founded: 1876

Mission:
The Appalachian Mountain Club promotes the protection, enjoyment, and wise use of the mountains, rivers, and trails of the Northeast.

Programs:
Programs include outdoor recreation, conservation, and environmental education. In addition, a monthly membership magazine and many books are produced, including trail guides, maps, and history books.

Atmosphere: Casual and hectic

Volunteers: Hundreds

STAFF POSITIONS
Full-time staff: 95 to 100   **Part-time staff:** Varies
Staff positions available: Varies
Staff salary: Low pay, great benefits

INTERNSHIPS
How Many: 10 or more   **Duration:** Varies
Remuneration: Unpaid
Duties:
Varies
Qualifications:
Interest in environmental issues.

How to apply: Cover letter and resume

JOBS
YOU CAN
LIVE WITH
1996

Arizona Toxics Information

Address: PO Box 1896
City: Bisbee   **State:** AZ   **Zip:** 85603

Telephone: (520) 432-5374   **Fax:** (520) 432-5374
E-mail: aztoxic@primenet.com
URL: Not listed

Other locations: None listed

Founded: 1990

Mission:
Advocacy for full public disclosure, public participation and pollution prevention strategies, in order to help bring about changes in hazardous materials policies and management that lead to the highest achievable levels of protection for public, occupational, and environmental health and to long-term sustainability of cultural and natural resources.

Programs:
1. Networking to facilitate the flow of information, encourage groups with different focuses, purposes, resources, and influence to recognize common goals and strategies in regard to toxics.
2. Enhancing the understanding of issues among activists, the public, and decisionmakers through research, maintenance, and development of our toxics database and clearinghouse, and mailings of informational packets on issues of concern.
3. Research information needed for furthering our mission.
4. Developing papers and articles on pertinent issues; providing testimony and comment to legislative and regulatory bodies.

Atmosphere: Casual, at times hectic

Volunteers: Varies

STAFF POSITIONS
Full-time staff: 3   **Part-time staff:** 2
Staff positions available: Varies
Staff salary: Depends on funding situation

INTERNSHIPS
How Many: 2 per year   **Duration:** Varies
Remuneration: Unpaid
Duties:
Office work, research assignments, individual projects.
Qualifications:
Ability to speak Spanish and some background in environmental issues.

How to apply: Cover letter, resume, application; phone calls welcome

Arms Control Association

Address: 1726 M Street, NW, Suite 201
City: Washington   **State:** DC   **Zip:** 20036

Telephone: (202) 463-8270   **Fax:** (202) 463-8273
E-mail: armscontrol@igc.apc.org
URL: http://www.igc.apc.org/aca

Other locations: None listed

Founded: 1971

Mission:
The Arms Control Association is a nonprofit, nonpartisan organization dedicated to increasing public understanding of the contribution arms control can make to national security.

Programs:
The Arms Control Association's programs help promote public understanding and support for effective arms control policies. Through its media and public education programs and its monthly magazine *Arms Control Today*, the Arms Control Association provides Congress, the media, and the interested public with authoritative information and analyses on arms control proposals, negotiations and agreements, and a broad range of related national security issues.

Atmosphere: Busy, informal, and casual

Volunteers: Varies

STAFF POSITIONS
Full-time staff: 11   **Part-time staff:** 1
Staff positions available: Varies
Staff salary: Varies

INTERNSHIPS
How Many: Varies   **Duration:** Semester or summer
Remuneration: $5 per day plus travel costs
Duties:
Interns research and write about national security issues; assist in preparing and editing the journal *Arms Control Today*; communicate with the press, government, and other arms control groups; monitor activity on Capitol Hill; and perform administrative tasks.

Qualifications:
Will consider undergraduates, recent graduates, and individuals no longer in school. Substantive background useful but less important than a willingness to work hard and learn.

How to apply: Cover letter, resume, application, and writing sample

Ashoka: Innovators for the Public Interest

Address: 1700 N. Moore Street, Suite 1920
City: Arlington   **State:** VA   **Zip:** 22209

Telephone: (703) 527-8300   **Fax:** (703) 527-8383
E-mail: Ashoka@tmn.com
URL: Not listed

Other locations: Chapter offices across the US and Europe

Founded: 1980

Mission:
Ashoka seeks to launch "public entrepreneurs" who have innovative ideas for change in developing countries. These entrepreneurs are dedicated to pattern-setting transformation in areas such as environment, education, women's issues, and health.

Programs:
Ashoka operates in 33 countries and has approximately 600 "fellows" in various fields. The Global Fellowship program links fellows working in the same areas, which translate to the Human-Environment Interface and All Children Learning programs. The Entrepreneur-to-Entrepreneur program links business people in the developed world with social entrepreneurs in the developing world. Major publications include *Leading Public* and *Entrepreneur*.

Atmosphere: Hectic, with a young multicultural staff

Volunteers: 5

STAFF POSITIONS
Full-time staff: 15   **Part-time staff:** 20
Staff positions available: 3 to 5 per year
Staff salary: Varies

INTERNSHIPS
How Many: Varies   **Duration:** A semester
Remuneration: Unpaid
Duties:
Depends on qualifications and educational experience.
Qualifications:
Responsibility, interest in developing countries, and strong analytical skills.

How to apply: Cover letter, resume and phone calls welcome

Association for Women in Science (AWIS)

Address: 1522 K Street, NW, Suite 820
City: Washington **State:** DC **Zip:** 20005

Telephone: (202) 408-0742 **Fax:** (202) 408-8321
E-mail: awis@awis.org
URL: Not listed

Other locations: None listed

Founded: 1971

Mission:
The Association for Women in Science works to achieve equity and full participation for women in science and technology.

Programs:
Two current programs are the AWIS community-based mentorship project, funded by the National Science Foundation, and developing a model program offering workable options to institutions committed to enhancing the academic climate for women science faculty, called "Women Scientists in Academia: Warming up a Chilly Climate," funded by the Alfred P. Sloan Foundation. There are also publications that encourage the participation of women and girls in science by promoting science education in schools and in the community.

Atmosphere: Relaxed

Volunteers: 1

STAFF POSITIONS
Full-time staff: 5 **Part-time staff:** 2
Staff positions available: 1 per year

Staff salary: Salary varies by position and degree of skill

INTERNSHIPS
How Many: 5 **Duration:** 2 to 4 months
Remuneration: $100 per week stipend
Duties:
Working closely with staff member, legislative analysis, research, writing for a magazine, composing press releases, and monitoring chapter activities.

Qualifications:
Interest in gender equity and science issues, competency in word processing, teamwork, and communication skills.

How to apply: Cover letter and resume

Association to Unite the Democracies

Address: 1506 Pennsylvania Avenue, SE
City: Washington   **State:** DC   **Zip:** 20003

Telephone: (202) 544-5150   **Fax:** (202) 544-3742
E-mail: AtUnite@aol.com
URL: Not listed

Other locations: None listed

Founded: 1940

Mission:
The Association to Unite the Democracies seeks the federation of the world's experienced democracies as the most effective means of addressing their common problems as well as those of mankind in general: freedom, peace and security, human rights, self-determination, commerce and trade, economic growth, overpopulation, the environment, and conservation of the world's finite resources.

Programs:
The Association awards Frank Fund Scholarships to graduate students interested in international federalism. Many of the Association's programs involve educating congressmen on international federalism and other related issues. The Association also publishes a newsletter titled *Unite!!*

Atmosphere: Casual and relaxed

Volunteers: Many

STAFF POSITIONS
Full-time staff: 1   **Part-time staff:** 3
Staff positions available: 2 per year
Staff salary: Varies

INTERNSHIPS
How Many: 3   **Duration:** Varies
Remuneration: Unpaid
Duties:
Research, Congressional liaison, and work with the Internet and World Wide Web.
Qualifications:
Interns must have an interest in international affairs and the issue of federalism, and possess computer skills, especially the Internet and World Wide Web.

How to apply: Cover letter and resume

JOBS YOU CAN LIVE WITH 1996

Atlantic Council of the United States

Address: 910 17th Street, NW, Suite 1000
City: Washington   **State:** DC   **Zip:** 20006

Telephone: (202) 463-4163   **Fax:** (202) 463-7241
E-mail: info@acgate.acus.org
URL: Not listed

Other locations: None listed

Founded: 1961

Mission:
The Council is a national, nonpartisan, nonprofit public policy center addressing US global interests in Atlantic and Pacific communities. The Council engages the US executive and legislative branches; the national and international business community, academe, and media; and diplomats and other foreign leaders. Programs foster informed public debate about foreign security and international economic interests and policies, and identify challenges and opportunities.

Programs:
Atlantic cooperation; NATO and European security; Atlantic and Pacific interrelationships; regional European issues; energy, environment and nuclear nonproliferation; and international security. Specific programs include: the Harriman Chair for East-West Studies; Office of Education and Outreach; and Office of Members and Finance.

Atmosphere: Relaxed, business-like, and hectic at times

Volunteers: 20

STAFF POSITIONS
Full-time staff: 30   **Part-time staff:** 0
Staff positions available: Varies
Staff salary: $21,500 per year plus health, vacation, and retirement

INTERNSHIPS
How Many: 15 to 20 per season   **Duration:** 8 to 12 weeks
Remuneration: Unpaid
Duties:
Involvement in developing, planning, and executing various programs and projects.
Qualifications:
WordPerfect 5.1, self-guided research, and organizational and communication skills.

How to apply: Cover letter, resume, academic transcript, writing sample, and two recommendations

Benton Foundation

Address: 1634 Eye Street, NW
City: Washington **State:** DC **Zip:** 20006

Telephone: (202) 638-5770 **Fax:** (202) 638-5771
E-mail: benton@benton.org
URL: http://cdinet.com/benton

Other locations: None listed

Founded: 1981

Mission:
The Benton Foundation works with nonprofit organizations to gain an effective voice for social change and to shape the emerging communications environment. The Foundation's efforts range from hands-on support for grassroots media advocacy to the "Public Interest Summit," a national forum for administration officials and 700 public-interest leaders to address the social, economic, and political impacts of the National Information Infrastructure.

Programs:
Strategic Communications—*Strategic Communications for Nonprofits*, a series of 10 guides to mass media, networking, and media production. "Advocacy Video Conference" with producers and media strategists from 17 countries to review and set standards for video use in social activism.
Children—helped create Coalition for America's Children, "Who's for Kids and Who's Just Kidding?" campaign, and "Children First" with Capital Cities/ABC-TV; helped develop and produce the Campaign to End Childhood Hunger, the Contract with America's Children, and national projects on health, education, safety, and security.

Atmosphere: Hectic and casual

Volunteers: 2

STAFF POSITIONS
Full-time staff: 9 **Part-time staff:** 1
Staff positions available: 1 per year
Staff salary: Depends on position and funding

INTERNSHIPS
How Many: As needed **Duration:** Varies
Remuneration: Varies
Duties:
Varies on specific program of interest.
Qualifications:
Experience in program area.
How to apply: Phone call

Bering Sea Coalition

Address: 730 I Street, Suite 200
City: Anchorage   **State:** AK   **Zip:** 99501

Telephone: (907) 279-6566   **Fax:** (907) 279-6228
E-mail: 1merculieff@igc.apc.org
URL: Not listed

Other locations: St. Paul Island, AK and Alberta, Canada

Founded: 1990

Mission:
The Bering Sea Coalition's mission is to be an advocate for health of the Bering Sea ecosystem and the viability of coastal indigenous cultures dependent on the Bering Sea.

Programs:
Work has focused on: four one-hour interlocking documentaries, working with indigenous peoples on the Russian and US side of the Bering Sea, helping indigenous cultures in rapid and involuntary transition, assisting with graduate research and lectures. The Coalition has also produced one book on island ecology.

Atmosphere: Casual
Volunteers: 0

STAFF POSITIONS
Full-time staff: 2   **Part-time staff:** 1
Staff positions available: Depends on program and funding
Staff salary: Varies

INTERNSHIPS
How Many: 2 to 9   **Duration:** 3 to 6 months
Remuneration: Stipend and travel expenses

Duties:
Research on issues dealing with fisheries, protecting ecosystems, indigenous rights, communication technology, and civilian uses of military technology.

Qualifications:
Good working knowledge of computer to be used for information transfer and communications, excellent research skills, willingness to explore new paradigms, and good writing skills.

How to apply: Cover letter, resume, academic transcript, writing sample, and written discussion of aspirations

Bigelow Laboratory for Ocean Sciences

Address: PO Box 475
City: West Boothbay Harbor   **State:** ME   **Zip:** 04575

Telephone: (207) 633-9600   **Fax:** (207) 633-9641
E-mail: Not listed
URL: http://www.bigelow.org/

Other locations: Affiliate of the University of New England in Bradford, Maine

Founded: 1974

Mission:
Bigelow Laboratory is an independent research institution where scientists conduct basic research on the biological, chemical, and physical processes that determine the productivity of the oceans. The Laboratory participates in educational and public service programs through affiliations with other organizations in the state of Maine and in the region.

Programs:
The organization works with federally-funded research programs, with National Science Foundation, Office of Naval Research, National Aeronautics and Space Administration, and the National Oceanographic and Atmosphere Administration. Scientists from the organization also publish papers regularly on scientific journals.

Atmosphere: Causal, hectic

Volunteers: 2

STAFF POSITIONS
Full-time staff: 40   **Part-time staff:** 0
Staff positions available: 1 to 2 per year
Staff salary: Varies

INTERNSHIPS
How Many: Varies   **Duration:** Varies
Remuneration: Currently unpaid
Duties:
Varies, depending on project.

Qualifications:
An interest and knowledge of related issues.

How to apply: Phone call and cover letter

JOBS YOU CAN LIVE WITH 1996

Blue Ridge Environmental Defense League, Inc.

Address: PO Box 88
City: Glendale Springs **State:** NC **Zip:** 28629

Telephone: (910) 982-2691 **Fax:** (910) 982-2954
E-mail: Not listed
URL: Not listed

Other locations: Wadesboro, NC; Marshall, NC; and Cherokee, NC

Founded: 1984

Mission:
The Blue Ridge Environmental Defense League's mission is threefold: Earth stewardship, public health, and environmental justice. The organization advocates grassroots action to empower whole communities on environmental issues. The staff provides assistance with research, training, strategy, and organization.

Programs:
Citizens' Nuclear Waste Watch, Earth Stage programs, Rural Appalachia Project, Cherokee Wake Up! Project, Family Farms Preservation Project, Clean Air Campaign, Citizens' Hazardous and Solid Waste Project. Publications include *Blue Ridge Environmental Times*, a quarterly newsletter; *Between the Lines*, issue updates; and bulletins titled *Waste Watcher*.

Atmosphere: Casual, home-like atmosphere

Volunteers: 200

STAFF POSITIONS
Full-time staff: 3 **Part-time staff:** 2
Staff positions available: Varies
Staff salary: $13,000 per year to start

INTERNSHIPS
How Many: 1 **Duration:** Varies
Remuneration: Unpaid
Duties:
Full range of staff activities.

Qualifications:
Flexible, self-reliant, personable.

How to apply: Cover letter, resume, and writing sample

Bread for the World

Address: 1100 Wayne Avenue, Suite 1000
City: Silver Spring **State:** MD **Zip:** 20910

Telephone: (301) 608-2400 **Fax:** (301) 608-2401
E-mail: bread@igc.org
URL: Not listed

Other locations: Chicago, IL; Minneapolis, MN; and Pasadena, CA

Founded: 1974

Mission:
Bread for the World is a nationwide Christian movement that seeks justice for the world's hungry people by lobbying our nation's decision makers.

Programs:
BFW's annual campaign is an "Offering of Letters," through which people are invited to contact their representatives concerning specific hunger-related legislation. The Bread for the World Institute produces an annual *Hunger Report*, which reports on the state of world hunger.

Atmosphere: Casual, hectic, and collegial

Volunteers: 7

STAFF POSITIONS
Full-time staff: 43 **Part-time staff:** 6
Staff positions available: Varies
Staff salary: Varies

INTERNSHIPS
How Many: Varies **Duration:** Varies
Remuneration: $14,000, health insurance possible
Duties:
Intern positions are available in administration, communications, development, issues, organizing, and research on hunger issues.

Qualifications:
An interest in public policy issues and a desire to make a difference in ending world hunger on domestic and international levels.

How to apply: Cover letter, resume, application, transcript, writing sample, and 3 recommendations

Break Away: The Alternative Break Connection

Address: 6026 Station B
City: Nashville   **State:** TN   **Zip:** 37235

Telephone: (615) 343-0385   **Fax:** (615) 343-3255
E-mail: BRAKAWAY@ctrvax.vanderbilt.edu
URL: Not listed

Other locations: None listed

Founded: 1991

Mission:
Break Away's mission is to promote service on the local, regional, national and international levels through break-oriented programs that immerse students in often different cultures, heighten social awareness, and advocate lifelong social action.

Programs:
Break Away: the Alternative Break Connection offers two basic services to their constituents. First, Break Away provides key information on planning and running a quality break program to colleges and high schools in their network through trainings and manuals. Second, Break Away helps place volunteers with community organizations through SiteBank, a database of volunteer opportunities both nationally and internationally.

Atmosphere: Casual and hectic

Volunteers: 0

STAFF POSITIONS
Full-time staff: 6   **Part-time staff:** 0
Staff positions available: 1 to 2 per year
Staff salary: $18,000 per year plus generous benefits

INTERNSHIPS
How Many: 3 to 4 per summer   **Duration:** 10 weeks
Remuneration: Stipend or academic credit

Duties:
Internships vary from year to year according to the needs of the organization. Past interns have revised manuals, organized special events, and recruited sites for their SiteBank.

Qualifications:
Experience in alternative break programs, eagerness to be a team player, ability to work independently, strong writing and communications skills, and computer experience.

How to apply: Application

JOBS YOU CAN LIVE WITH 1996

British American Security Information Council (BASIC)

Address: 1900 L Street, NW, Suite 401
City: Washington **State:** DC **Zip:** 20036

Telephone: (202) 785-1266 **Fax:** (202) 387-6298
E-mail: Not listed
URL: http://www.igc.apc.org/basic/

Other locations: London, England

Founded: 1987

Mission:
BASIC analyzes international security policy in Europe and North America; promotes public awareness of defense, disarmament, military strategy, and nuclear policies to foster informed debate on these issues; facilitates the exchange of information and analysis among researchers, journalists, and parliamentarians on both sides of the Atlantic; encourages decisionmakers to take advantage of emerging opportunities for disarmament and cooperation.

Programs:
Major programs include nuclear disarmament and nonproliferation, arms trade issues, and European security. Major publications are *BASIC Reports*, *BASIC Papers*, and *BASIC Notes*.

Atmosphere: Semiprofessional, relaxed, friendly, and busy

Volunteers: 0

STAFF POSITIONS
Full-time staff: 7 **Part-time staff:** 0
Staff positions available: 1 per year
Staff salary: $20,000 per year to start

INTERNSHIPS
How Many: 3 **Duration:** 4 months
Remuneration: Varies

Duties:
Substantial policy research and some administrative responsibilities, including answering phones, clipping newspapers, maintaining the Peacenet account, conducting research on the World Wide Web, obtaining publications.

Qualifications:
Experience using software applications, especially Word Perfect and File Maker Pro, a general knowledge of how nonprofits work, and an interest in peace and international security.

How to apply: Cover letter, resume, academic transcript, writing sample, and two letters of recommendation

Campaign for UN Reform

Address: 713 D Street, SE
City: Washington   **State:** DC   **Zip:** 20003-0270

Telephone: (202) 546-3956   **Fax:** (202) 328-2126
E-mail: Not listed
URL: Not listed

Other locations: None listed

Founded: 1975

Mission:
The Campaign is a bipartisan organization with a primary goal to gain support for an enhanced, reformed, restructured UN. The Campaign advances this mission through public education, lobbying, and electioneering.

Programs:
The programs include a 14-point plan for UN reform to: create an international disarmament organization, improve peacekeeping capability, improve human rights machinery, establish special international criminal court, enhance environment and conservation programs, increase use of the International Court of Justice, modify veto in Security Council, improve mechanisms to settle disputes, reform administrative system, improve General Assembly decisionmaking, provide adequate and stable revenues, restructure world trade and monetary system, consolidate development programs, and create programs for areas not under national control.

Atmosphere: Casual, serious

Volunteers: 10 to 20

STAFF POSITIONS
Full-time staff: 1   **Part-time staff:** Varies
Staff positions available: None
Staff salary: Varies

INTERNSHIPS
How Many: 1 to 2   **Duration:** Usually a semester
Remuneration: $50 per week for travel
Duties:
Professional duties such as writing press releases and representing the organization at various conferences and meetings.
Qualifications:
Strong interest in political activity directed toward global structures for peace.

How to apply: Phone call

Campus Green Vote

Address: 1731 Connecticut Avenue, NW, Suite 501
City: Washington   **State:** DC   **Zip:** 20009

Telephone: (202) 234-5990   **Fax:** (202) 234-5997
E-mail: cgv@igc.apc.org
URL: http://www.cgv.org/cgv

Other locations: None listed

Founded: 1992

Mission:
Campus Green Vote is dedicated to training and organizing a diverse, national network of young voters to protect the environment.

Programs:
Youth Vote '96—Campus Green Vote is spearheading this coalition that is committed to registering, educating, and mobilizing 12 million young people to go to the polls this November.
Campus Green Pages—a directory of youth and campus environmental leaders nationwide.
Activist Profile Network—a component on the World Wide Web where student campaigns are highlighted.
Campus Green Vote—field operation nationwide to register, educate, and mobilize young people to the polls to vote for the environment.

Atmosphere: Casual, fast-paced, and productive campaign atmosphere

Volunteers: 20

STAFF POSITIONS
Full-time staff: 4   **Part-time staff:** 2
Staff positions available: Varies

Staff salary: Teens to low twenties per year

INTERNSHIPS
How Many: Unlimited   **Duration:** Varies
Remuneration: College credit, possible wage
Duties:
Based on interest and includes everything from writing updates and working on the Internet to student organizing and field support.

Qualifications:
Commitment to the oganization's mission, energy and enthusiasm, and familiarity or experience in related areas.

How to apply: Cover letter, resume, application, and writing sample; phone calls welcome

Campus Outreach Opportunity League (COOL)

Address: 1511 K Street, NW, Suite 307
City: Washington **State:** DC **Zip:** 20005

Telephone: (202) 637-7004 **Fax:** (202) 637-7031
E-mail: COOL2YOU1@aol.com
URL: Not listed

Other locations: None listed

Founded: 1984

Mission:
COOL works to educate and empower college students to strengthen our nation through community service.

Programs:
COOL provides technical assistance and consultation to emerging campus-based community service programs, as well as to already established programs. COOL also sponsors national programs such as: "Into the Streets," a one-day trial service program; produces publications, newsletters and books; COOL Leaders, student development program; and a national conference that attracts over 2000 students every year.

Atmosphere: Casual, efficient

Volunteers: 2

STAFF POSITIONS
Full-time staff: 4   **Part-time staff:** 8
Staff positions available: 1 per year
Staff salary: Varies

INTERNSHIPS
How Many: 10 per year   **Duration:** 3 to 8 months
Remuneration: Some, amount varies
Duties:
Interns work on research and development of new program initiatives.

Qualifications:
Desire to work on social change efforts, experience or interest in community service.

How to apply: Cover letter and resume; phone calls welcome

Career Development Group (CDG)

Address: 40 Wall Street, Suite 2124
City: New York   **State:** NY   **Zip:** 10005-1301

Telephone: (212) 759-2368   **Fax:** (212) 793-5723
E-mail: npc@igc.apc.org
URL: Not listed

Other locations: Chicago, IL; Los Angeles, CA; Vancouver, BC; and Mongolia

Founded: 1991

Mission:
CDG's mission is to help nonprofit organizations, educational institutions, and government agencies become more effective and self-sufficient through appropriate governance, direction, management, and use of technology.

Programs:
CDG and its parent organization, Non-Profit Computing Inc., has provided free services to nonprofit organizations, educational institutions, and government agencies as a direct services provider and as a provider of information and referral. These services include: donations of equipment and software, pro-bono volunteer consulting, internships and apprenticeships, job readiness training and job networking groups, sponsorship and co-sponsorship of user support groups, networks, conferences, and other events.

Atmosphere: Varies, as work takes place at client sites

Volunteers: Hundreds

STAFF POSITIONS
Full-time staff: 1   **Part-time staff:** 3
Staff positions available: Varies
Staff salary: All are volunteers, but this may change.

INTERNSHIPS
How Many: As many as possible   **Duration:** Varies
Remuneration: Depends on specific client
Duties:
Depends on what is needed on a specific project.
Qualifications:
Some computer literacy and an interest in nonprofit work.
How to apply: Cover letter and resume

Carrying Capacity Network

Address: 2000 P Street, NW, Suite 240
City: Washington   **State:** DC   **Zip:** 20036

Telephone: (202) 296-4548   **Fax:** (202) 296-4609
E-mail: CCN@igc.apc.org
URL: Not listed

Other locations: None listed

Founded: 1989

Mission:
Carrying Capacity Network's action-oriented initiatives focus on achieving national revitalization, population stabilization, immigration reduction, sustainable economies, and resource conservation to preserve our heritage and quality of life.

Programs:
Carrying Capacity Network's programs include:
Network Bulletin—a bimonthly newsletter that serves as a primary mobilization vehicle, communication link, and information resource for the Network.
FOCUS—a quarterly journal that includes in-depth articles and interviews.
Fax Alert System—helps keep Network members updated on fast-breaking developments.

Atmosphere: Small, casual office

Volunteers: 0

STAFF POSITIONS
Full-time staff: 5   **Part-time staff:** 1
Staff positions available: 2 to 3 per year
Staff salary: Varies

INTERNSHIPS
How Many: 2   **Duration:** Variable
Remuneration: Paid and unpaid available
Duties:
Mostly administrative: phone, mail, some research and writing.
Qualifications:
Excellent writing skills and an interest in Carrying Capacity issues.

How to apply: Cover letter and resume; phone calls welcome

Carter Center

Address: One Copenhill
City: Atlanta   **State:** GA   **Zip:** 30307

Telephone: (404) 331-3900   **Fax:** (404) 331-0283
E-mail: Not listed
URL: http://www.emory.edu/CARTER_CENTER

Other locations: None listed

Founded: 1982

Mission:
The Carter Center brings people and resources together to resolve conflicts, promote peace and human rights, and fight disease, hunger, poverty, and oppression. The Center works through collaborative efforts in three areas: international democratization and development, global health, and urban revitalization.

Programs:
The Carter Center's programs are divided into three areas:
Democratization and Development—this includes African Governance Program, Commission on Radio and Television Policy, Conflict Resolution Program, Global Development Program, Human Rights Program, and the Latin American and Caribbean Program.
Global Health—agriculture, Guinea Worm Eradication Program, Interfaith Health Program, Mental Health Program, Task Force for Child Survival and Development.
Urban Revitalization—the Atlanta Project and the America Project.

Atmosphere: Busy

Volunteers: 130

STAFF POSITIONS
Full-time staff: 250   **Part-time staff:** 50
Staff positions available: Varies

Staff salary: Varies

INTERNSHIPS
How Many: Varies   **Duration:** Varies
Remuneration: Unpaid
Duties:
Interns are given a broad range of duties focusing on issues addressed by their program, also office administration and issues cutting across programs.

Qualifications:
Superior academic ability and course work, professional or personal experience, and career interests related to Carter Center programs.

How to apply: Cover letter, resume, application, transcripts, and letters of recommendation

Center for Campus Organizing

Address: PO Box 748
City: Cambridge   **State:** MA   **Zip:** 02139

Telephone: (617) 354-9363   **Fax:** (617) 547-5067
E-mail: ucp@igc.apc.org
URL: http://envirolink.org/orga/xucp.

Other locations: None listed

Founded: 1991

Mission:
Center for Campus Organizing serves as a clearinghouse and network for progressive student activists and alternative journals and newspapers.

Programs:
Provides tools for action and education. The Center has a national campus contact and alternative journalism network where they connect discussion lists on e-mail. They also coordinated the National Days of Campus Action Against the Contract with America.

Atmosphere: Casual, ideal for college students and recent graduates

Volunteers: Many

STAFF POSITIONS
Full-time staff: 4   **Part-time staff:** 2
Staff positions available: 2 to 3 per year
Staff salary: $18,000 per year to start

INTERNSHIPS
How Many: Unlimited   **Duration:** 2 to 6 months
Remuneration: Stipend available
Duties:
Office work, research, campus networking, e-mail communication.

Qualifications:
An interest and history in student activism, volunteer or political work, research skills.

How to apply: Cover letter, resume, and application; phone calls welcome

Center for Clean Air Policy

Address: 444 N. Capitol Street, Suite 602
City: Washington **State:** DC **Zip:** 20001

Telephone: (202) 624-7709 **Fax:** (202) 508-3829
E-mail: Not listed
URL: Not listed

Other locations: Prague, Czech Republic

Founded: 1985

Mission:
The Center's work is guided by the belief that sound energy and environmental policy solutions serve both economic and environmental interests. The Center's work emphasizes the need for cost-effective, pragmatic, and holistic, long-term solutions to energy and environmental problems that provide win-win situations for everyone involved.

Programs:
The Center's programs include: Southeast Wisconsin Dialogue on Ozone Pollution and Global Climate Change; Small Engine Trade-In Pilot Program in Wisconsin; Air Quality and Electricity Restructuring; Transatlantic Collaboration to Improve Transportation, Land Use and Air Quality; The Offsets Forum; US/Costa Rica Climate Change Initiative; and German Marshall Fund Environmental Fellowship Program. The Center also offers a number of publications on related environmental issues.

Atmosphere: Casual, relaxed, can be hectic

Volunteers: 0

STAFF POSITIONS
Full-time staff: 9   **Part-time staff:** 2
Staff positions available: 2 to 3 per year
Staff salary: Varies

INTERNSHIPS
How Many: 1 to 2 a semester   **Duration:** 1 semester
Remuneration: Unpaid
Duties:
Administrative duties as well as research, depending on experience and ability.
Qualifications:
Bright and interested in market-based approaches to environmental issues. Strong writing skills, and the ability to work alone or with others.

How to apply: Cover letter, resume, and writing sample

Center for Defense Information (CDI)

Address: 1500 Massachusetts Avenue, NW
City: Washington **State:** DC **Zip:** 20005

Telephone: (202) 862-0700 **Fax:** (202) 862-0708
E-mail: jhazen@cdi.org
URL: Not listed

Other locations: None listed

Founded: 1972

Mission:
CDI believes in a strong economy, an open government, an educated public, a healthy environment, and a strong military for the nation's security. CDI opposes excessive expenditures for weapons and policies that increase the dangers of war. CDI was founded to serve as an independent monitor of the military and is regarded as the foremost research organization in the country analyzing military spending, policies, and weapons systems.

Programs:
CDI has outreach channels to educate the public and Congress on military matters, including the award-winning journal *The Defense Monitor*, *America's Defense Monitor* on PBS and cable (an advisory council of more than 100 retired senior military officers), and a syndicated radio show titled *The Question of the Week*. CDI works to inform the media on related issues and has dealt with all major networks, CNN, MacNeil/Lehrer Newshour, and radio and television broadcasts to Sweden, Great Britain, Canada, Australia, Germany, and Japan. CDI maintains an extensive library of research materials used by journalists, investigators, Congress, and the public.

Atmosphere: Relaxed, casual, busy

Volunteers: 3

STAFF POSITIONS
Full-time staff: 25 **Part-time staff:** 0
Staff positions available: 1 to 3 per year
Staff salary: Varies

INTERNSHIPS
How Many: 4 per trimester **Duration:** Varies
Remuneration: $700 monthly stipend
Duties:
Research, administrative, and television production associates.
Qualifications:
An interest in broadcast communications, military issues, and public policy are encouraged but not required. High academic achievements and writing skills are essential.

How to apply: Resume, transcript, two letters of recommendation, writing sample, letter describing interests and goals

Center for Democracy and Technology

Address: 1001 G Street, NW, #500E
City: Washington   **State:** DC   **Zip:** 20001

Telephone: (202) 637-9800   **Fax:** (202) 637-0896
E-mail: ask@cdt.org
URL: http://www.cdt.org

Other locations: None listed

Founded: 1994

Mission:
The Center for Democracy and Technology seeks to develop and implement public policies that preserve and enhance constitutional civil liberties and democratic values in new interactive media.

Programs:
The Center coordinates several policy working groups in issues such as industry and the public interest, free speech and expression in interactive media, communications privacy and security, and confidentiality of personal data. A monthly newsletter titled *CDT Policy Roots* is published, as are various position papers and reports.

Atmosphere: Casual, focused, hard-working

Volunteers: 0

STAFF POSITIONS
Full-time staff: 7   **Part-time staff:** 1
Staff positions available: 1 to 2 per year
Staff salary: Varies

INTERNSHIPS
How Many: 1 to 3   **Duration:** Varies
Remuneration: Sometimes
Duties:
Depends on experience, interest, and abilities.

Qualifications:
Intelligence, dedication to civil liberties issues, interest in communications, and a sense of humor.

How to apply: Cover letter and resume; phone calls welcome

Center for Economic Conversion (CEC)

Address: 222 View Street
City: Mountain View   **State:** CA   **Zip:** 94041

Telephone: (415) 968-8798   **Fax:** (415) 968-1126
E-mail: cec@igc.org
URL: http://www.conversion.org

Other locations: None listed

Founded: 1975

Mission:
CEC is a nonprofit, nonpartisan, public-benefit corporation dedicated to building a sustainable, peace-oriented economy. The Center works at national, state, and local levels to reduce economic dependence on excessive military spending and reinvest resources to meet human and environmental needs.

Programs:
CEC takes part in research, public education, and advocacy through print, broadcast, and electronic media, including the publication of the newsletter *Positive Alternatives*. CEC also works with conversion planning and implementation by identifying, monitoring, and promoting notable conversion efforts, creating and fostering conversion demonstration projects, and assisting groups in base closure planning.

Atmosphere: Professional, friendly, supportive, casual, dedicated

Volunteers: 7

STAFF POSITIONS
Full-time staff: 3   **Part-time staff:** 2
Staff positions available: 0 to 1 per year
Staff salary: Competitive salary with excellent benefits

INTERNSHIPS
How Many: 6   **Duration:** Flexible, min. 3 months
Remuneration: Unpaid
Duties:
Depends on specific projects.

Qualifications:
Well organized and efficient; work independently; Macintosh literate; familiar with the Internet, World Wide Web, library research procedures; and committed to positive social change.

How to apply: Cover letter and resume

Center for Global Change

Address: 7100 Baltimore Avenue, Suite 401
City: College Park **State:** MD **Zip:** 20740

Telephone: (301) 403-4165 **Fax:** (301) 403-4292
E-mail: Not listed
URL: Not listed

Other locations: Washington, DC

Founded: 1989

Mission:
The Center seeks innovative solutions to global environmental problems; studies the relationship of energy use, equity and economic development; evaluates and recommends technology, policy, and institutional reforms promoting sustainable development and reducing the risk of environmental degradation, particularly those posed by climate change and ozone depletion. Research is disseminated through publications, seminars, training, and outreach.

Programs:
Global Change—a journal designed to provide accurate information about scientific and economic research on problems of global environmental change and their implications for public policy and private investment decisions.
Renewable Energy Institute—purpose is to accelerate the worldwide development and use of renewable energy technologies.
Training Program on the Climate Change Convention—created a resource package for conducting national and local workshops in developing countries on climate change and the UN Framework Convention on Climate Change.

Atmosphere: Casual and hectic

Volunteers: 0

STAFF POSITIONS
Full-time staff: 6 **Part-time staff:** 7
Staff positions available: 0 to 1 per year
Staff salary: Not listed

INTERNSHIPS
How Many: Not listed **Duration:** Not listed
Remuneration: Not listed
Duties:
Not listed
Qualifications:
Not listed
How to apply: Not listed

Center for Integrated Agricultural Systems

Address: 1450 Linden Drive, Room 146
City: Madison   **State:** WI   **Zip:** 53706

Telephone: (608) 262-5200   **Fax:** (608) 265-3020
E-mail: Not listed
URL: Not listed

Other locations: None listed

Founded: 1989

Mission:
The primary purpose of this organization is to support farmers in their efforts to increase profitability, improve their family's quality of life, and protect natural resources. Using an interdisciplinary approach, the Center examines systems of agriculture, natural resources, economic organizations, and human communities, while emphasizing sustainability.

Programs:
Programs involve grazing dairy systems, food systems, community-supported agriculture, farmer networks, and economic analysis of sustainable agricultural practices.

Atmosphere: Casual and very productive

Volunteers: 0

STAFF POSITIONS
Full-time staff: 1   **Part-time staff:** 3
Staff positions available: 1 to 2 per year
Staff salary: $30,000 to $35,000 per year with benefits for project coordinators

INTERNSHIPS
How Many: Varies   **Duration:** Varies
Remuneration: Varies
Duties:
Depends on specific projects.

Qualifications:
Enthusiasm, interest in issues, dependability, and willingness to learn.

How to apply: Cover letter and resume

Center for Marine Conservation

Address: 1725 DeSales Street, NW, Suite 500
City: Washington **State:** DC **Zip:** 20036

Telephone: (202) 429-5609 **Fax:** (202) 872-0619
E-mail: Not listed
URL: Not listed

Other locations: Hampton, VA; St. Petersburg, FL; Marathon, FL; and San Francisco, CA
Founded: 1972

Mission:
The Center for Marine Conservation is a private, nonprofit environmental organization dedicated to protection of the marine environment and its wildlife, specifically marine mammals, sea turtles, fishes, and their respective habitats. The Center accomplishes this goal through public awareness and education, advocacy, grassroots activities, and collaborative efforts with others, both nationally and internationally.

Programs:
Programs include Marine Debris and Entanglement Program, Sea Turtle Conservation Program, Marine Habitat Conservation Program, Marine Mammal Conservation Program, Fisheries Conservation Program, and Marine Biological Diversity Project. The Center also produces three newsletters, *Coastal Connections*, *Marine Conservation News*, and *Sanctuary Currents*.

Atmosphere: Casual

Volunteers: 1

STAFF POSITIONS
Full-time staff: 53 **Part-time staff:** 2 to 3
Staff positions available: Varies
Staff salary: Varies

INTERNSHIPS
How Many: 1 to 3 **Duration:** Minimum 3 months
Remuneration: Commuting expenses and academic credit
Duties:
Internships are available in communications and finance, as well as all program areas.

Qualifications:
Enthusiasm, initiative, responsibility, and an ability to write well.

How to apply: Phone call

Center for Policy Alternatives

Address: 1875 Connecticut Avenue, NW, Suite 710
City: Washington **State:** DC **Zip:** 20009

Telephone: (202) 387-6030 **Fax:** (202) 986-2539
E-mail: CFPA@CAPACCESS.ORG
URL: Not listed

Other locations: None listed

Founded: 1975

Mission:
The Center for Policy Alternatives is a nonprofit organization that promotes progressive public policy across all 50 states. A nonpartisan think tank, CPA connects innovative people and ideas to build a new economy which is inclusive, sustainable, and just. CPA's elected officials and activists promote pragmatic change to support families, strengthen communities, conserve resources for future generations, and enhance democratic participation by every citizen.

Programs:
Mobilizing Participation for Tomorrow's Communities—empower citizens by encouraging collaborative solutions.
Creating Community Capital—bring together policymakers, investors, activists to increase capital to underfinanced sectors.
Women's Voices for the Economy—unite women to move the economy forward.
State-Federal Relations—balancing responsibilities around trade and unfunded mandates to ensure fundamental protections.
Leadership—through Policy Alternatives Leaders program, State Issues Forum, and Flemming Fellows Leadership Institute.

Atmosphere: Casual dress and hard work

Volunteers: 3

STAFF POSITIONS
Full-time staff: 22 **Part-time staff:** 2
Staff positions available: 3 to 4 per year

Staff salary: Not listed

INTERNSHIPS
How Many: 30 per year **Duration:** Varies
Remuneration: Varies
Duties:
Research, writing, and analysis.

Qualifications:
Academic achievement; excellent research, analysis and writing skills; commitment to issue areas.

How to apply: Cover letter, resume, and writing sample

Center for Psychology and Social Change

Address: 1493 Cambridge Street
City: Cambridge **State:** MA **Zip:** 02139

Telephone: (617) 497-1553 **Fax:** (617) 497-0122
E-mail: Not listed
URL: Not listed

Other locations: None listed

Founded: 1983

Mission:
The Center for Psychology and Social Change is an important contributor in seeking to discover what it really means to be fully human, sensing that we have somehow become disconnected from a deeper source of knowing and being. The Center attracts and supports individuals with the creative energy and courage to break out of our culture's dominant ways of thinking, and the discipline and skill to share their insights in useful and reapplicable ways.

Programs:
The Center's major programs involve eco-psychology, political development of children, and conflict resolution. The Center also publishes a magazine titled *Centerpiece*.

Atmosphere: Casual

Volunteers: 6

STAFF POSITIONS
Full-time staff: 4 **Part-time staff:** 0
Staff positions available: 1 per year
Staff salary: Varies

INTERNSHIPS
How Many: 4 **Duration:** Usually a semester
Remuneration: College credit, small stipend
Duties:
Varies with position.

Qualifications:
Good skills in communication, organization, and computers.

How to apply: Cover letter, resume, and writing sample

Center for Rural Affairs

Address: PO Box 406
City: Walthill   **State:** NE   **Zip:** 68067-0406

Telephone: (402) 846-5428   **Fax:** (402) 846-5420
E-mail: HN1721@HANDSNET.ORG
URL: Not listed

Other locations: Hartington, NE

Founded: 1973

Mission:
The Center is committed to building sustainable rural communities consistent with social and economic justice, stewardship of the natural environment, and broad distribution of wealth. The goal is to advance the vision of rural America through research, education, advocacy, organizing, and leadership development.

Programs:
Community-based loan programs; linking beginning farmers with landowners; beginning farmer advocacy, education, and support network; on-farm research; agricultural research policy; federal agricultural credit policy; federal farm bill and conservation legislation; state corporate farming bills; land-grant and federal agricultural research; Nebraska property tax reform and school financing; numerous publications and newsletters.

Atmosphere: Casual, relaxed, hectic schedules.

Volunteers: 0

STAFF POSITIONS
Full-time staff: 22   **Part-time staff:** 2
Staff positions available: 1 per year
Staff salary: $22,000 to $26,000 per year plus benefits

INTERNSHIPS
How Many: 0   **Duration:** Not listed
Remuneration: Not listed
Duties:
Not listed
Qualifications:
Not listed
How to apply: Not listed

Center for Science in the Public Interest

Address: 1875 Connecticut Avenue, NW, #300
City: Washington **State:** DC **Zip:** 20009

Telephone: (202) 332-9110 **Fax:** (202) 265-4954
E-mail: KenH@essential.org
URL: Not listed

Other locations: None listed

Founded: 1971

Mission:
The Center for Science in the Public Interest is a consumer advocacy organization whose twin missions are to conduct innovative research and advocacy programs in health and nutrition, and to provide consumers with current, useful information about their health and well-being.

Programs:
The Center's programs involve: providing useful, objective information to the public and policy makers; conducting research on food, alcohol, health, the environment, and other issues related to science and technology; representing the citizen's interests before regulatory, judicial and legislative bodies on food, alcohol, health, the environment, and other issues; ensuring that science and technology are used for the public good; and encouraging scientists to engage in public-interest activities.

Atmosphere: Casual, relaxed, hectic

Volunteers: 5

STAFF POSITIONS
Full-time staff: 43 **Part-time staff:** 5
Staff positions available: 5 per year
Staff salary: $19,000 to $21,000 per year to start

INTERNSHIPS
How Many: Varies **Duration:** 10 weeks
Remuneration: $4.25 to $5.25 per hour
Duties:
Varies depending on specific project and area of interest.

Qualifications:
Strong writing and computer skills, strong academic background in related areas, a science background is useful but not required.

How to apply: Cover letter, resume, academic transcript, writing sample, and two letters of recommendation

Center for the Study of Ethics in the Professions (CSEP)

Address: 3101 S. Dearborn, Room 166
City: Chicago **State:** IL **Zip:** 60616

Telephone: (312) 567-3017 **Fax:** (312) 567-3016
E-mail: Weil@charlie.acc.iit.edu
URL: http://www.iit.edu/ncsep

Other locations: None listed

Founded: 1976

Mission:
The Center works to advance teaching and research in professional ethics. Engineering and science are two professions on which the Center has concentrated.

Programs:
Research, curriculum building, empirical projects in scientific decisionmaking and secrecy in science, development of similar programs for professional societies, production of a number of publications, and publication of other works through grants.

Atmosphere: Comfortable, busy, serious, hectic

Volunteers: Varies

STAFF POSITIONS
Full-time staff: 4 **Part-time staff:** 2
Staff positions available: Varies
Staff salary: Varies

INTERNSHIPS
How Many: Varies **Duration:** Flexible
Remuneration: Unpaid
Duties:
Library research, participation in proposal initiation and development, editing periodicals.

Qualifications:
Intelligence, energy, adaptability, and good humor.

How to apply: Cover letter and resume; phone calls welcome

JOBS YOU CAN LIVE WITH 1996

Center for War, Peace and the News Media

Address: 10 Washington Place, 4th Floor
City: New York **State:** NY **Zip:** 10003

Telephone: (212) 998-7618 **Fax:** (212) 995-4143
E-mail: cwpnm@acfcluster.nyu.edu
URL: Not listed

Other locations: One office in Boston, MA and five throughout Russia

Founded: 1985

Mission:
The Center, a nonprofit, nonpartisan organization devoted to the study and improvement of news reporting of international security issues, focuses on aiding American journalists in coverage of international affairs and assisting in the development of a professional, economically sound, and democratic media in Russia. The Center organizes briefings and seminars, and publishes papers and briefs especially for journalists covering these issues.

Programs:
Russian-American Press and Information Center (RAPIC)—contributes to development of independent, effective, and financially-solvent Russian media as a pillar of democracy.
European Security Network (ESN)—works with editors and producers to assist them with coverage of European issues.
International Non-Proliferation Media Project—enhances reporting of nuclear proliferation and related post-Cold War security issues.
Pacific Rim Security Network—offers US and Asian journalists programs similar to the ESN.

Atmosphere: Fast-paced, highly organized, professional/friendly staff

Volunteers: 3

STAFF POSITIONS
Full-time staff: 40 **Part-time staff:** 5
Staff positions available: Varies

Staff salary: Varies

INTERNSHIPS
How Many: 3 **Duration:** Semester or year
Remuneration: School credit
Duties:
Research, writing reports, logistical support, and general project assistance.
Qualifications:
Interns should be graduate or advanced undergraduate students with knowledge of media, security studies, international relations, conflict resolution, and/or Russia.

How to apply: Cover letter, resume, and writing samples

Center for Women Policy Studies

Address: 2000 P Street, NW, Suite 508
City: Washington   **State:** DC   **Zip:** 20036

Telephone: (202) 872-1770   **Fax:** (202) 296-8962
E-mail: Not listed
URL: Not listed

Other locations: None listed

Founded: 1972

Mission:
The Center for Women Policy Studies is a multiethnic feminist organization that serves as an independent policy research and advocacy institution.

Programs:
The Center's programs address: educational equity, work/family and workplace diversity policies, economic opportunity for low-income women, violence against women and girls, women's health, reproductive rights and health, women and AIDS, and leadership development.

Atmosphere: Busy, casual

Volunteers: 2 to 3

STAFF POSITIONS
Full-time staff: 10   **Part-time staff:** 0
Staff positions available: 1 to 2 per year
Staff salary: Varies

INTERNSHIPS
How Many: 2 to 3   **Duration:** 2 to 3 months
Remuneration: Academic credit only
Duties:
Research, writing, and administrative.

Qualifications:
Demonstrated interest in women's issues.

How to apply: Cover letter and resume

JOBS YOU CAN LIVE WITH 1996

Chesapeake Bay Foundation

Address: 162 Prince George Street
City: Annapolis   **State:** MD   **Zip:** 21401

Telephone: (410) 268-8816   **Fax:** (410) 268-6687
E-mail: Not listed
URL: Not listed

Other locations: Harrisburg, PA and Richmond, VA

Founded: 1966

Mission:
The Foundation seeks to restore and sustain the bay's ecosystem by substantially improving the water quality and productivity of the watershed with respect to water clarity, resilience of the system, and diversity and abundance of its living resources, and to maintain a high quality of life for the people of the Chesapeake Bay region.

Programs:
The Chesapeake Bay Foundation's programs revolve around environmental education, resource protection, and land conservation. The Chesapeake Bay Foundation also publishes an annual report and the *CBF Newsletter*.

Atmosphere: Casual, hectic, and dedicated

Volunteers: 200

STAFF POSITIONS
Full-time staff: 130   **Part-time staff:** 7
Staff positions available: 5 per year
Staff salary: Varies

INTERNSHIPS
How Many: Varies   **Duration:** 3 months
Remuneration: Academic credit or stipend
Duties:
Varies according to intern interest and abilities.

Qualifications:
Hard working with a sincere commitment and dedication to the mission of the organization.

How to apply: Cover letter and resume

Citizens Clearinghouse for Hazardous Waste (CCHW)

Address: 119 Rowell Court, PO Box 6806
City: Falls Church   **State:** VA   **Zip:** 22040

Telephone: (703) 237-2249   **Fax:** (703) 237-8389
E-mail: cchw@essential.org
URL: Not listed

Other locations: None listed

Founded: 1981

Mission:
CCHW is a national, nonprofit organization that supports grassroots environmental groups by providing technical assistance and leadership development. Areas of focus include the protection of health and environment against corporate pollution and unjust environmental practices.

Programs:
CCHW has over 70 guidebooks on organizing and technical issues to assist community groups in informing and empowering themselves on health and environmental protection. The current major program is the Stop Dioxin Exposure Campaign.

Atmosphere: Casual, hectic, some relaxed times

Volunteers: Varies

STAFF POSITIONS
Full-time staff: 10   **Part-time staff:** 1
Staff positions available: Varies
Staff salary: Not listed

INTERNSHIPS
How Many: Varies   **Duration:** Varies
Remuneration: Travel expenses
Duties:
Research, translation of science into layman's terms, organizing, writing.
Qualifications:
Commitment to social justice and good writing skills.
How to apply: Cover letter and resume

Citizens for Alternatives to Chemical Contamination (CACC)

Address: 8735 Maple Grove Road
City: Lake   **State:** MI   **Zip:** 48632-9511

Telephone: (517) 544-3318   **Fax:** (517) 544-3318
E-mail: Not listed
URL: Not listed

Other locations: None listed

Founded: 1978

Mission:
CACC is a grassroots environmental educational and advocacy organization dedicated to the principles of social justice, citizen empowerment, pollution prevention, and the protection of the Great Lakes ecosystem.

Programs:
CACC holds the Backyard Eco Conference, which is an annual event that takes place in the spring. The CACC Clearinghouse newsletter is published 10 times per year.

Atmosphere: Casual

Volunteers: Varies

STAFF POSITIONS
Full-time staff: 1   **Part-time staff:** 1
Staff positions available: 0 to 1 per year
Staff salary: Depends on fundraising success

INTERNSHIPS
How Many: 1 to 2   **Duration:** 3 months, but flexible
Remuneration: Sometimes
Duties:
Depends on skills and organizational need.
Qualifications:
Flexibility, dedication to "the cause," imagination.

How to apply: Cover letter and resume

Clean Water Action

Address: 1320 18th Street, NW, Suite 300
City: Washington **State:** DC **Zip:** 20036

Telephone: (202) 457-1286 **Fax:** (202) 457-0287
E-mail: Not listed
URL: Not listed

Other locations: 18 offices across the country

Founded: 1973

Mission:
Clean Water Action is a national, nonprofit, grassroots lobbying organization working for preservation of water resources, quality drinking water, recycling, reduction of toxins, and a sustainable environment.

Programs:
CWA campaigns for "Green" candidates; fights to save our vanishing wetlands; stops the "Wise Use" movement; saves the Everglades (make big sugar pay); stops incinerators; promotes recycling; promotes "Buy Recycled" programs; promotes "Home Safe Home" programs—help pollution prevention to begin at home; protects watersheds from contamination; stops the destruction of groundwater resources; protects the Chesapeake and Delaware Bays; works to pass battery recycling legislation, mercury control, and lead abatement laws; and ensures that all people in every community receive environmental justice.

Atmosphere: Casual, relaxed, fun

Volunteers: Varies

STAFF POSITIONS
Full-time staff: 40 **Part-time staff:** 10 to 15
Staff positions available: Open ended
Staff salary: $17,000 to $20,000 per year with full benefits

INTERNSHIPS
How Many: Varies **Duration:** Varies
Remuneration: See staff salary
Duties:
Political organizers do direct work on political campaigns and policy campaigns, helping in passing legislation through grassroots lobbying.

Qualifications:
Strong communication skills and a motivation to help the environment.

How to apply: Cover letter and resume; phone calls welcome

JOBS YOU CAN LIVE WITH 1996

Co-op America

Address: 1612 K Street, NW, Suite 600
City: Washington   **State:** DC   **Zip:** 20006

Telephone: (202) 832-5307   **Fax:** (202) 331-8166
E-mail: ca@cals.com
URL: Not listed

Other locations: None listed

Founded: 1982

Mission:
Co-op America is a national, nonprofit organization that links socially responsible businesses and consumers in a national network, an alternative marketplace. Co-op America educates people about how to vote with their dollars for a more peaceful, just society and a healthier environment. Co-op America provides information, products, and services to help make spending and saving choices that work in harmony with politics and values.

Programs:
Co-op America educates people to use spending and investing power to bring values of social justice and environmental sustainability into the economy. They also assist socially and environmentally responsible businesses to emerge and thrive, and pressure irresponsible companies to adopt socially and environmentally responsible practices. Publications: *Co-op America Quarterly*; *Boycott Action News*; *National Green Pages*; *You, Your Money, and The World: Financial Planning for a Better Tomorrow*; and *Connections* (a newsletter for business members).

Atmosphere: A casual, progressive, cooperative environment

Volunteers: Varies

STAFF POSITIONS
Full-time staff: 20   **Part-time staff:** 6
Staff positions available: 1 to 2 per year
Staff salary: $19,000 per year plus benefits to start

INTERNSHIPS
How Many: Up to 20   **Duration:** Varies
Remuneration: $50 monthly, free membership
Duties:
Various types of internships are available, and the duties will vary.

Qualifications:
At least 2 years of college or equivalent experience.

How to apply: Cover letter and resume

Committee for the National Institute for the Environment (CNIE)

Address: 730 11th Street, NW, 3rd Floor
City: Washington **State:** DC **Zip:** 20001-4521

Telephone: (202) 628-4303 **Fax:** (202) 628-4311
E-mail: cnie@access.digex.net
URL: http://www.inhs.uiuc.edu/niewww/cnie.html

Other locations: None listed

Founded: 1990

Mission:
The National Institute for the Environment (NIE) will provide the information needed to solve our country's complex environmental problems. As an independent, nonregulatory, federal science institute, the NIE will improve the scientific basis for making environmental decisions.

Programs:
Programs involve outreach and education to scientific, environmental, and community groups, state and local governments, and businesses. Publications include an NIE newsletter, federal environmental research and development programs, and issue papers on media outreach and environmental science advocacy.

Atmosphere: Focused and relaxed

Volunteers: Many

STAFF POSITIONS
Full-time staff: 10   **Part-time staff:** 6
Staff positions available: 3 to 4 per year
Staff salary: Depends on individual circumstances

INTERNSHIPS
How Many: 2 to 3   **Duration:** 3 to 6 months
Remuneration: Sometimes
Duties:
Program support in areas of outreach, education, media, legislation, and research.
Qualifications:
Energy and maturity.

How to apply: Cover letter and resume

Common Cause

Address: 2030 M Street, NW, Suite 300
City: Washington   **State:** DC   **Zip:** 20036

Telephone: (202) 833-1200   **Fax:** (202) 659-3716
E-mail: Not listed
URL: Not listed

Other locations: None listed

Founded: 1970

Mission:
Common Cause is a nonprofit, nonpartisan citizens' lobby that is working to restore ethics in government and to curb the undue influence of lobbyists and special-interest money in government. Common Cause also works to protect the civil rights of all citizens and to make government more open and accountable.

Programs:
Common Cause magazine is a political journal which studies special-interest political contributions. Other programs vary but are related to their mission statement.

Atmosphere: Casual, very professional

Volunteers: 30

STAFF POSITIONS
Full-time staff: 45   **Part-time staff:** 2
Staff positions available: Varies
Staff salary: $19,000 per year and health benefits to start

INTERNSHIPS
How Many: Varies   **Duration:** A semester or summer
Remuneration: Commuting expenses
Duties:
Grassroots organizing, contributing research to policy papers and investigative studies, monitoring Congressional committee meetings, working as press office aide, researching for *Common Cause* magazine.

Qualifications:
Strong interest in related issues.

How to apply: Cover letter, application, writing sample, and two letters of reference

CONCERN, Inc.

Address: 1794 Columbia Road, NW
City: Washington   **State:** DC   **Zip:** 20009

Telephone: (202) 328-8160   **Fax:** (202) 387-3378
E-mail: concern@igc.apc.org
URL: Not listed

Other locations: None listed

Founded: 1970

Mission:
CONCERN seeks to broaden public participation in environmental protection and to promote informed action by supplying citizens with literature needed to be effective advocates for policies and programs that improve environmental quality and public health. CONCERN believes change begins with individuals, protection of the environment should be a core principle in community decisionmaking, and prevention should take precedence over remedial solutions.

Programs:
CONCERN provides individuals and organizations in private and public sectors with information on a broad range of issues; works with other organizations to achieve certain goals, such as the promotion of energy efficiency, conservation, and use of renewable fuels; develops new programs to help communities achieve environmental and economic health and social equity through comprehensive, inclusive, long-term planning; produces publications on issues such as pesticides, global warming, household waste, water safety, sustainable agriculture.

Atmosphere: Casual

Volunteers: 1 to 2

STAFF POSITIONS
Full-time staff: 2 to 8   **Part-time staff:** 2 to 3
Staff positions available: 0 to 1 per year
Staff salary: Varies

INTERNSHIPS
How Many: 2   **Duration:** Varies
Remuneration: $75 per week
Duties:
Interns assist with research, compile documentation on programs, help produce written materials, participate in community outreach programs, attend workshops, and assist with communications.

Qualifications:
Knowledge of environmental issues, strong research and writing abilities, good interpersonal communications skills, and word processing ability desirable.

How to apply: Cover letter, resume, and writing sample

Concord Feminist Health Center (CFHC)

Address: 38 S. Main Street
City: Concord   **State:** NH   **Zip:** 03301-4888

Telephone: (603) 225-2739   **Fax:** (603) 228-6255
E-mail: Not listed
URL: Not listed

Other locations: None listed

Founded: 1974

Mission:
CFHC is a nonprofit organization dedicated to reproductive rights and the health and well-being of women. The mission is to empower women through health-care services and information consistent with a feminist philosophy; to provide safe, legal abortion and gynecological care, health education, and referral information; to work politically in support for sexual and reproductive rights and social justice for women; and to treat every woman with respect.

Programs:
Pregnancy testing and counseling, first trimester abortion and follow-up, gynecological care, contraceptive counseling and care, STD and HIV testing, information and referrals for health care, community education on women's health issues, legislative lobbying, and the publication of *WomenWise*, a quarterly women's health journal.

Atmosphere: Casual, professional

Volunteers: 10

STAFF POSITIONS
Full-time staff: 10   **Part-time staff:** 8
Staff positions available: Varies, usually 1 to 2 per year
Staff salary: $8.75 per hour to start

INTERNSHIPS
How Many: 1 to 2 per year   **Duration:** 6 months to 1 year
Remuneration: Varies
Duties:
Writing/research for *WomenWise*, updating referral information, organizing.

Qualifications:
Pro-choice with a strong interest in women's health issues. An ability to be direct, compassionate, and to communicate clearly both verbally and in writing.

How to apply: Cover letter and resume; phone calls welcome

JOBS YOU CAN LIVE WITH 1996

Conservation Career Development Program (CCDP)

Address: 1800 N. Kent Street, Suite 1260
City: Arlington **State:** VA **Zip:** 22209

Telephone: (703) 524-2441 **Fax:** (703) 524-2451
E-mail: Not listed
URL: Not listed

Other locations: Denver, CO; Los Angeles, CA; Newark, NJ; Oakland, CA; and Seattle, WA
Founded: 1990

Mission:
Under the auspices of the Student Conservation Association, CCDP seeks to develop the "green" workforce of tomorrow by striving to diversify the environment/conservation pool. CCDP provides opportunities for and encourages the participation of those traditionally underrepresented in such fields, namely, women, people of color, and persons with disabilities; yet CCDP is by no means exclusive of others.

Programs:
CCDP High School Program, CCDP Fellows Program, SCA/CCDP Career Services, Youth Forest Camp

Atmosphere: Casual, businesslike

Volunteers: 450

STAFF POSITIONS
Full-time staff: About 20 **Part-time staff:** 2 to 3
Staff positions available: 2 per year
Staff salary: Varies

INTERNSHIPS
How Many: Varies **Duration:** 2 to 4 months
Remuneration: Benefits vary
Duties:
Varies with program.

Qualifications:
An interest in related issues.

How to apply: Cover letter, resume, application, academic transcript, writing sample, references, and an interview

Conservation International

Address: 1015 18th Street, NW, Suite 1000
City: Washington **State:** DC **Zip:** 20036

Telephone: (202) 429-5660 **Fax:** (202) 887-5188
E-mail: J.McLaughlin@conservation.org
URL: http://www.conservation.org

Other locations: Portland, OR; Peru; Costa Rica; Bolivia; Ecuador; Manila; Botswana; Papua, New Guinea; Jakarta; and Guatemala
Founded: 1987

Mission:
Conservation International is dedicated to the protection of natural ecosystems and the species that rely on these habitats for survival.

Programs:
Programs focus on: ecosystem conservation, training local communities, small enterprise, research, biological inventories, and environmental education.

Atmosphere: Casual, hectic

Volunteers: 5

STAFF POSITIONS
Full-time staff: 160 **Part-time staff:** 25
Staff positions available: 5 to 10 per year
Staff salary: Varies on education and experience

INTERNSHIPS
How Many: 20 per year **Duration:** 1 month to 1 year
Remuneration: Up to $7.50 hourly or $1500 monthly
Duties:
Research, writing, filing, and mailing.

Qualifications:
Interest in conservation, ability to speak foreign languages, and interest in foreign travel.

How to apply: Phone call

Consumer Energy Council of America Research Foundation (CECA/RF)

Address: 2000 L Street, NW, Suite 802
City: Washington **State:** DC **Zip:** 20036

Telephone: (202) 659-0404 **Fax:** (202) 659-0407
E-mail: ceca1@aol.com
URL: Not listed

Other locations: None listed

Founded: 1973

Mission:
CECA/RF is a nonprofit organization with a primary commitment to the provision of reliable and affordable energy for all sectors of the nation, with special regard to environmental consequences. CECA/RF places great emphasis on conducting impartial research on energy issues and building coalitions between the public and private sector.

Programs:
CECA/RF focuses on the economic and social impacts of energy policies, and provides technical assistance to public and private sector organizations. CECA/RF has a unique and highly respected expertise in creating forums for consensus-building on public policy issues among public and private sector organizations, state and local organizations, businesses, utilities, consumer and environmental organizations, and government agencies.

Atmosphere: Casual

Volunteers:

STAFF POSITIONS
Full-time staff: 5 to 10 **Part-time staff:** 2 to 3
Staff positions available: 2 to 3 per year
Staff salary: Varies

INTERNSHIPS
How Many: 2 to 4 **Duration:** 8 weeks to 1 year
Remuneration: Academic credit, some stipends

Duties:
Interns will assist in basic research, drafting issue papers, conducting literature reviews, analyzing data, editing, legislative duties, and proofreading.

Qualifications:
Interns must have excellent relevant academic credentials as well as good research and writing skills. Ability to work independently and experience with WordPerfect 5.1 are helpful.

How to apply: Cover letter, resume, three references, and two in-depth writing samples

JOBS YOU CAN LIVE WITH 1996

Cosanti Foundation

Address: HC 74 Box 4136
City: Mayer   **State:** AZ   **Zip:** 86333

Telephone: (520) 632-7135   **Fax:** (520) 632-6229
E-mail: arcosanti@aol.com
URL: http://www.arcosanti.org

Other locations: Scottsdale, AZ

Founded: 1965

Mission:
The mission of the Cosanti Foundation is the continuing research into arcology—a study which combines the discipline of architecture and ecology—through educational outreach of the continuing construction of Arcosanti, a prototype arcology.

Programs:
Programs include Arcosanti workshops; silt, clay, and bronze workshop; and various educational programs related to arcology and the construction of Arcosanti, the prototype project.

Atmosphere: Casual, some construction work

Volunteers: 10 to 20 per month

STAFF POSITIONS
Full-time staff: 70   **Part-time staff:** 10
Staff positions available: 20 per year
Staff salary: Minimum wage with paid holidays, and medical/dental insurance

INTERNSHIPS
How Many: Varies   **Duration:** Varies
Remuneration: Varies

Duties:
Duties could range from public relations and outreach to construction and architectural drafting.

Qualifications:
Flexibility and a willingness to contribute to the project.

How to apply: Cover letter, resume, and application

Council For A Livable World

Address: 110 Maryland Avenue, NE, Suite 409
City: Washington   **State:** DC   **Zip:** 20002

Telephone: (202) 543-4100   **Fax:** (202) 543-6297
E-mail: jsmith@clw.org
URL: http://www.clw.org/pub/clw/welcome.html

Other locations: None listed

Founded: 1962

Mission:
The Council was founded by Leo Szilard, who recognized the importance of controlling the nuclear menace and felt an organization was needed to influence and educate Congress and the public on the dangers of nuclear war. The primary activities are lobbying Congress on nuclear arms control issues and being a political-action committee that raises money for candidates that have shown a commitment to reducing nuclear weapons stockpiles.

Programs:
Programs and publications revolve around the issues of the military industrial complex, nuclear issues, military budget issues, conventional arms trade, and United Nations peacekeeping issues. The Council also produces a quarterly newsletter that features similar issues. The Council for a Livable World Education Fund was formed in 1980 to finance the research, resource production, and dissemination of information components of various projects in order to educate the public and policy makers on peacekeeping, the UN, and conventional arms transfers.

Atmosphere: Fun and fast-paced

Volunteers: 3

STAFF POSITIONS
Full-time staff: 12   **Part-time staff:** 1
Staff positions available: 0 to 1 per year
Staff salary: Varies

INTERNSHIPS
How Many: 2   **Duration:** 3 to 6 months
Remuneration: Commuting expenses
Duties:
Depends on specific project.

Qualifications:
A history of interest in UN issues, controlling the arms race, global arms sales, national spending priorities, and general political subjects.

How to apply: Cover letter, resume, and writing sample; phone calls welcome

JOBS
YOU CAN
LIVE WITH
1996

Council for Responsible Genetics

Address: 5 Upland Road, Suite 3
City: Cambridge   **State:** MA   **Zip:** 02140

Telephone: (617) 868-0870   **Fax:** (617) 491-5344
E-mail: gwatch@neu.edu
URL: Not listed

Other locations: None listed

Founded: 1983

Mission:
The Council for Responsible Genetics educates the public about the social and environmental implications of genetic technologies and advocates for developments in biotechnology that serve the public interest.

Programs:
Activities focus on:
- Countering discrimination on the basis of predictive genetic tests.
- Strengthening environmental regulation of commercial biotechnology.
- Opposing the patenting of genetically engineered life forms.
- Maintaining an information clearinghouse open to the public.
- Publishing *GeneWATCH*, the Center's journal dedicated to monitoring the social and environmental implications of biotechnology.

Atmosphere: Exciting and fast-paced

Volunteers: 3

STAFF POSITIONS
Full-time staff: 3   **Part-time staff:** 0
Staff positions available: 1 per year
Staff salary: Commensurate with experience

INTERNSHIPS
How Many: 3   **Duration:** Varies
Remuneration: Unpaid
Duties:
Research, writing analysis, outreach.

Qualifications:
Quick learner, able to work with little supervision, science background is not required.

How to apply: Cover letter and resume; no phone calls, please

JOBS YOU CAN LIVE WITH 1996

Council on Economic Priorities

Address: 30 Irving Place
City: New York   **State:** NY   **Zip:** 10003

Telephone: (212) 420-1133   **Fax:** (212) 420-0988
E-mail: cep@echo.nyc.com
URL: Not listed

Other locations: None listed

Founded: 1969

Mission:
The Council on Economic Priorities is dedicated to the impartial and accurate analysis of the social and environmental record of corporations. CEP provides information to consumers, investors, and employers, and provides the incentive for superior corporate social responsibility.

Programs:
CEP rates companies on: the environment, women's advancement, minority advancement, disclosure, charitable giving, community outreach, family benefits, and workplace issues. CEP projects, dedicated to finding the truth about corporate behavior, bringing about reform, and enhancing superior corporate social and environmental performance, include: *Shopping for a Better World: The Quick and Easy Guide to ALL Your Socially Responsible Shopping, SCREEN, Global SCREEN,* Project on International Security, Campaign for Cleaner Corporations, America's Corporate Conscience Awards, and Project on Transnational Corporations.

Atmosphere: Casual, young, energetic staff

Volunteers: Varies

STAFF POSITIONS
Full-time staff: 13   **Part-time staff:** 3
Staff positions available: 2 to 3 per year
Staff salary: $18,000 to $21,000 per year to start

INTERNSHIPS
How Many: 13 to 15   **Duration:** 10 weeks
Remuneration: $165 per week
Duties:
A wide variety, from research to administrative duties.

Qualifications:
A background in environmental studies, economics, or international relations.

How to apply: Cover letter, resume, and writing sample; phone calls welcome (Mid-March deadline for summer internships)

JOBS YOU CAN LIVE WITH 1996

Council on Hemispheric Affairs (COHA)

Address: 724 9th Street, NW, Suite 401
City: Washington **State:** DC **Zip:** 20001

Telephone: (202) 393-3322 **Fax:** (202) 393-3423
E-mail: COHA@csa.deltal.org
URL: Not listed

Other locations: None listed

Founded: 1975

Mission:
COHA analyzes, researches, and publishes materials on the social, economic, and political issues associated with Inter-American relations. Issues include fair trade rather than free trade, human rights and social justice issues, monitoring excesses in US regional policy, and promoting democratic procedures in Latin America. Recently, COHA has been critical of Clinton administration policy, opposed NAFTA, and advocated normalizing relations with Cuba.

Programs:
COHA publishes a highly esteemed professional publication, the *Washington Report on the Hemisphere,* a biweekly offered by subscription. It also distributes one or two lengthy research memoranda per week to the media and policy makers, relating to breaking issues. COHA research findings are frequently placed in *The Congressional Record* and COHA researchers have the opportunity of seeing their op-ed articles placed in newspapers throughout the country. COHA researchers also are called upon by radio and television producers to present their analysis to a wide range of audiences.

Atmosphere: Relaxed, casual

Volunteers: 25

STAFF POSITIONS
Full-time staff: 3 **Part-time staff:** 7
Staff positions available: 2 per year
Staff salary: Varies

INTERNSHIPS
How Many: 25 **Duration:** 3 to 4 months
Remuneration: All volunteers
Duties:
Interns are expected to research, write, and distribute their findings. All interns take turns as managing editor of COHA's biweekly publication.

Qualifications:
Dedication, flexible, able to take on responsibility, and to bring their very best to the organization.

How to apply: Cover letter, resume, application, academic transcript, and writing sample; phone calls welcome

JOBS YOU CAN LIVE WITH 1996

Cultural Survival, Inc.

Address: 46 Brattle Street
City: Cambridge   **State:** MA   **Zip:** 02138

Telephone: (617) 441-5400 x641   **Fax:** (617) 441-5417
E-mail: survival@husc.harvard.edu
URL: Not listed

Other locations: None listed

Founded: 1972

Mission:
Cultural Survival, Inc. is dedicated to preserving, promoting, and protecting the human rights of indigenous peoples and ethnic minorities throughout the world—primarily through land demarcation.

Programs:
Special programs vary, but one long-term project is Cultural Survival Enterprises, which is an alternative trading internship program. They also produce the *Cultural Survival Quarterly*, an academic journal dealing with related issues.

Atmosphere: Varies

Volunteers: 20

STAFF POSITIONS
Full-time staff: 6   **Part-time staff:** 2
Staff positions available: 1 to 2 per year
Staff salary: Varies

INTERNSHIPS
How Many: Over 20   **Duration:** 12 weeks or more
Remuneration: Unpaid
Duties:
 Hands-on assistance in any department.
Qualifications:
Bright, self-starter, with an anthro/social/human rights interest.

How to apply: Cover letter, resume, and application; phone calls welcome

JOBS YOU CAN LIVE WITH 1996

Dakota Resource Council

Address: PO Box 1095
City: Dickinson   **State:** ND   **Zip:** 58602-1095

Telephone: (701) 227-1851   **Fax:** (701) 225-8315
E-mail: Not listed
URL: Not listed

Other locations: None listed

Founded: 1978

Mission:
The Dakota Resource Council's mission is to use community organizing to help self-sustaining local citizen groups participate in public policy decisionmaking that affects their lives. The Council places an emphasis on protecting natural resources and economic justice for family farmers. The majority of their work is centered in western North Dakota.

Programs:
Major programs include:
Family Farm Preservation Campaign—farm justice issues.
Don't Waste the West—preventing industrial-scale waste disposal in North Dakota.
Law and Order in the Cornfields—watchdog activities concerning state and federal strip-mining laws.
Protect Wind Harvest—promoting the development of North Dakota's greatest sustainable energy resource.
Corporate Responsibility Campaign—battling legislation and administrative practices that limit citizen control on corporate action.

Atmosphere: Casual

Volunteers: Varies

STAFF POSITIONS
Full-time staff: 4   **Part-time staff:** 1
Staff positions available: 1 per year
Staff salary: $15,000 per year plus full benefits to start

INTERNSHIPS
How Many: 1   **Duration:** 8 to 10 weeks
Remuneration: Negotiable
Duties:
Interns are usually hired for a previously discussed, specific project.
Qualifications:
Self-starter, good communication skills, and a commitment to organization's goals.

How to apply: Resume and writing sample; phone calls welcome

Dakota Rural Action (DRA)

Address: Box 549
City: Brookings   **State:** SD   **Zip:** 57006

Telephone: (605) 697-5204   **Fax:** (605) 697-6230
E-mail: Not listed
URL: Not listed

Other locations: Two offices in Brookings, SD

Founded: 1987

Mission:
Dakota Rural Action, a grassroots organization dedicated to social and economic justice for rural America, has chapters on issues of local concern; has changed local, state, and national policies affecting family farm and ranch agriculture, waste reduction, recycling, toxic waste reduction, and tax justice; works on community organizing, leadership development, action research, and holding public officials accountable; and produces publications.

Programs:
Corporate Accountability in Agriculture—designed to help stop further corporate domination and exploitation of our food system, with emphasis on stopping meat packer concentration.
Family Farm and Ranch Preservation Project—dedicated to sound stewardship and sustainable agricultural practices; responsible rural credit and conservation programs; and ensuring beginning farm and ranch opportunities and existing family-sized producers stay on the land.
Don't Waste the West Campaign—helps stop importation of waste to SD and promotes corporate responsibility in waste industry.

Atmosphere: Busy, hectic, usually casual

Volunteers: 0

STAFF POSITIONS
Full-time staff: 4   **Part-time staff:** 2
Staff positions available: About 1 every year

Staff salary: $14,000 to $16,000 per year, depending on experience

INTERNSHIPS
How Many: 2 per year   **Duration:** 3 to 12 months
Remuneration: Stipend, room, board, travel
Duties:
Research, writing, lay-out, fundraising, and some general organizing.

Qualifications:
Interest and/or experience in community organizing.

How to apply: Cover letter, resume, and writing sample

Defenders of Wildlife

Address: 1101 14th Street, NW, Suite 1400
City: Washington   **State:** DC   **Zip:** 20005

Telephone: (202) 682-9400   **Fax:** (202) 682-1331
E-mail: Not listed
URL: http://www.defenders.org

Other locations: Oregon, Alaska, Arizona, Florida, Montana, and New Mexico

Founded: 1947

Mission:
Dedicated to the protection of all native wild animals and plants in their natural habitats, Defenders of Wildlife focuses on what scientists consider two of the most serious environmental threats to the planet: the accelerating rate of extinction of species and associated loss of biological diversity, and habitat alteration and destruction. Defenders of Wildlife advocates new approaches to conservation that will help keep species from becoming endangered.

Programs:
Species Conservation—endangered species recovery, predator protection and restoration, wild bird protection, and marine mammal protection.
Habitat Conservation—ecosystem protection, refuge reform, gap analysis, marine biodiversity, *Watchable Wildlife* programs and books, programs that encourage protection of ecosystems and interconnected habitats.
Policy Leadership—Endangered Species Act, state and federal wildlife laws, biodiversity education, wildlife trade, international treaties.

Atmosphere: Casual, but hectic at times

Volunteers: 2

STAFF POSITIONS
Full-time staff: 55   **Part-time staff:** 3
Staff positions available: Depends on funding and turnover

Staff salary: $19,000 to $23,000 per year plus benefits

INTERNSHIPS
How Many: Looking for more   **Duration:** 3 months
Remuneration: Possible $500 per month
Duties:
Depends on specific area of expertise.

Qualifications:
Varies, but a strong interest and/or knowledge of environmental issues is important.

How to apply: Cover letter, resume, and writing sample

Delaware Valley Citizens' Council For Clean Air (Clean Air Council)

Address: 135 S. 19th Street, Suite 300
City: Philadelphia **State:** PA **Zip:** 19103

Telephone: (215) 567-4004 **Fax:** (215) 567-5791
E-mail: office@cleanair.org
URL: http://www.libertynet.org/~clean air/

Other locations: Harrisburg, PA

Founded: 1967

Mission:
The mission of the Council is to inform and educate the public concerning the health, economic, and aesthetic effects of air pollution and the technological and legal tools available for its control; to promote understanding of the role of national, state, and local governments in controlling pollution; and to stimulate efforts to clean the air, including a mandatory curbside recycling law and aggressive waste reduction program.

Programs:
Transportation—promotes clean options, such as mass transit, pedestrian-friendliness, low-emission vehicles, and bike paths.
Energy—works to restructure the electric utility industry to reduce air pollution.
Clean Air Act Implementation—uses media outreach, constituency building, and litigation to ensure that Pennsylvania is meeting its obligations under the Clean Air Act.
Environmental Problems in Low-Income Communities—outreach on issues like local pollution, childhood lead poisoning and asthma, and contaminated or abandoned industrial sites.

Atmosphere: Casual, relaxed, hectic

Volunteers: 9

STAFF POSITIONS
Full-time staff: 7 **Part-time staff:** 13
Staff positions available: 1 to 3 per year
Staff salary: Negotiable

INTERNSHIPS
How Many: Varies **Duration:** One semester
Remuneration: Negotiable
Duties:
Research, writing, field work, organizing, and clerical work.
Qualifications:
Interest in the environment, good organizational and writing skills, self-motivated, creative, able to work well with others.

How to apply: Phone call

Demilitarization for Democracy (DFD)

Address: 2001 S Street, NW, Suite 630
City: Washington **State:** DC **Zip:** 20009

Telephone: (202) 319-7191 **Fax:** (202) 319-7194/0937
E-mail: pdd@clark.net
URL: Not listed

Other locations: None listed

Founded: 1992

Mission:
DFD, a nonprofit research and advocacy center that works to change US policy toward the world from one of militarization to one of demilitarization. DFD believes that the goals of demilitarization and democracy are linked and that a dramatic reduction in the size and political power of armed forces in developing countries will spur progress toward full democratic rights.

Programs:
The Code of Conduct Campaign—a nationwide, long-term effort to ban arms sales to dictators, human rights abusers, countries at war, and countries not cooperating with international arms control efforts.
The International Campaign to Ban Landmines.
The Year 2000 Campaign to Redirect World Military Spending to Human Rights Development—publications are available on this topic.

Atmosphere: Small, relaxed staff that works well under pressure

Volunteers: 0

STAFF POSITIONS
Full-time staff: 5 **Part-time staff:** 1
Staff positions available: Usually one per year

Staff salary: Varies, but includes full health/life insurance

INTERNSHIPS
How Many: 2 **Duration:** Usually 6 months
Remuneration: Unpaid
Duties:
Assisting the director on major campaigns, research, representing DFD in coalitions and at meetings.

Qualifications:
Commitment, enthusiasm, hard working, and interested.

How to apply: Cover letter, resume, and a 2-page writing sample

Dodge Nature Center

Address: 1795 Charlton Street
City: West St. Paul **State:** MN **Zip:** 55107

Telephone: (612) 455-4531 **Fax:** (612) 455-2175
E-mail: Not listed
URL: Not listed

Other locations: None listed

Founded: 1967

Mission:
The Dodge Nature Center works to provide environmental education to kindergarten through sixth grade students.

Programs:
The Dodge Nature Center's programs revolve around implementing an environmental kindergarten-to-sixth grade curriculum, with activities on: Earth Day, Spring Arrivals Festival, Fall Harvest Festival, family public programs, and producing a quarterly newsletter.

Atmosphere: Casual, relaxed, hectic at times

Volunteers: 20

STAFF POSITIONS
Full-time staff: 15 **Part-time staff:** 7 to 8
Staff positions available: Varies
Staff salary: Varies

INTERNSHIPS
How Many: 3 to 4 per season **Duration:** 10 to 12 weeks
Remuneration: $100 per week
Duties:
Extensive teaching opportunities; care of reptiles, amphibians, and raptors; gardening assistance; evening and weekend program development; and special projects.

Qualifications:
Motivation, enthusiasm, quick learner, teaching skills, enjoy working with children, and public contact.

How to apply: Cover letter, resume, and academic transcript

E: The Environmental Magazine

Address: PO Box 5098
City: Westport **State:** CT **Zip:** 06881

Telephone: (203) 854-5559 **Fax:** (203) 866-0602
E-mail: AXGM65A@PRODIGY.COM
URL: Not listed

Other locations: None listed

Founded: 1990

Mission:
E Magazine provides national and international coverage of environmental issues, serves as an independent voice for the environmental movement, and is a vital information source for anyone concerned about the environment.

Programs:
All programs revolve around the production of the magazine.

Atmosphere: A casual, dedicated work atmosphere

Volunteers: 4

STAFF POSITIONS
Full-time staff: 7 **Part-time staff:** 2
Staff positions available: Varies
Staff salary: Varies

INTERNSHIPS
How Many: 2 to 3 per semester **Duration:** 3 months
Remuneration: Academic credit
Duties:
Assist with writing, researching, fact-checking, proofreading layouts, summarizing articles for Eco-Net, and general office duties.

Qualifications:
Writing background preferred, strong communication skills, and a healthy interest in environmental issues.

How to apply: Cover letter, resume, and writing samples

Earth Island Institute

Address: 300 Broadway, #28
City: San Francisco   **State:** CA   **Zip:** 94110

Telephone: (415) 788-3666   **Fax:** (415) 788-7324
E-mail: earthisland@igc.apc.org
URL: http://www.earthisland.org/ei/

Other locations: None listed

Founded: 1982

Mission:
The Earth Island Institute seeks to develop innovative projects that support the conservation, preservation, and restoration of the earth.

Programs:
International Marine Mammal Project, Urban Habitat Project, Sea Turtle Restoration Project, Rethink Paper, Global Service Corps, Tibetan Plateau Project, *Earth Island Journal.*

Atmosphere: Casual, dedicated workaholics

Volunteers: 1 to 10

STAFF POSITIONS
Full-time staff: 30   **Part-time staff:** 10
Staff positions available: 0 to 1 per year
Staff salary: $16,000 to $24,000 per year

INTERNSHIPS
How Many: Varies   **Duration:** Varies
Remuneration: Unpaid
Duties:
Research, writing, administrative/clerical, merchandise, editing, Web page.
Qualifications:
Not listed.

How to apply: Cover letter, resume, and application

Earth Share

Address: 3400 International Drive, NW, Suite 2K
City: Washington **State:** DC **Zip:** 20008

Telephone: (800) 875-3863 **Fax:** (202) 537-7101
E-mail: Not listed
URL: http://www.earthshare.org

Other locations: Field representatives nationwide

Founded: 1989

Mission:
Earth Share is a federation of environmental and conservation charities. The goals are to expand the financial support and the involvement of individuals, primarily at the workplace, to advance affiliate organizations' efforts to promote and protect public health and welfare and to conserve natural resources for future generations.

Programs:
1. Provides free educational information, nationally on television and radio and in magazines and newspapers, with the goal of educating people about how they can protect the environment.
2. Works to increase participation of grassroots organizations in the national environmental movement, allowing greater racial, ethnic, and socioeconomic diversity represented by these organizations.
3. Provides information on its affiliate agencies to employees in payroll education charitable fundraising drives, increasing the effectiveness of their 43-member agencies in protecting the world's health and natural resources.

Atmosphere: Casual

Volunteers: Varies

STAFF POSITIONS
Full-time staff: 10 **Part-time staff:** 7
Staff positions available: Varies

Staff salary: Varies

INTERNSHIPS
How Many: Varies **Duration:** Flexible, 2 to 4 months
Remuneration: Unpaid
Duties:
Depends on specific program and intern level of experience.

Qualifications:
Interest in environmental issues, public speaking ability, and good organizational skills.

How to apply: Cover letter and resume

Earth Train

Address: 5 Spring Court
City: Orinda   **State:** CA   **Zip:** 94563

Telephone: (510) 254-9101   **Fax:** (510) 254-8728
E-mail: gonzo@well.com
URL: Not listed

Other locations: Overland Park, KS

Founded: 1990

Mission:
The Earth Train's mission is to establish a global network of millions of youth gaining the knowledge, experience, and support they need to create an environmentally sustainable and equitable world.

Programs:
Two-week and one-month expeditions to Puerto Rico, Mexico, and India for youth ages15 to 19. These expeditions focus on leadership skills, team building, communication, and community service through ecological studies. The Youth Action Agenda is an environmental education curriculum that works with regional schools to combine hands-on leadership activities with community service projects that are based on environmental principles. Earth Train also offers an environmental leadership retreat for students and teachers.

Atmosphere: Casual, relaxed, productive, and open

Volunteers: 3

STAFF POSITIONS
Full-time staff: 3   **Part-time staff:** 1
Staff positions available: Not listed
Staff salary: Varies

INTERNSHIPS
How Many: 3 to 6   **Duration:** School year term
Remuneration: $50 per week

Duties:
Visiting chapter schools, teaching curriculum once a week, giving presentations, working with teachers and students, retreat coordination, and group facilitation.

Qualifications:
Dependable, responsible, excellent communication skills, ability to work with a variety of people (including high school students), interest and knowledge of environmental issues, and flexibility.

How to apply: Cover letter, resume, and application; phone calls welcome

Earthtrust

Address: 25 Kaneohe Bay Drive, #205
City: Kailua **State:** HI **Zip:** 96734

Telephone: (808) 254-2866 **Fax:** (808) 254-6409
E-mail: earthtrust@mcimail.com
URL: http://baki.kcc.hawaii.edu/~et

Other locations: None listed

Founded: 1976

Mission:
Earthtrust's mission revolves around international wildlife conservation and establishes original, innovative programs in support of these issues.

Programs:
Anti-Whaling—has pioneered DNA research and analysis of whale meat products; on-going analysis of whale meat in Asia in order to achieve an end to "scientific whaling."
Driftnet Work—networking on a global scale to monitor driftnet activity and put pressure on governments to stop this fishery technique.

Atmosphere: Casual, hectic, serious, and very focused

Volunteers: Varies

STAFF POSITIONS
Full-time staff: Varies   **Part-time staff:** Varies
Staff positions available: Varies
Staff salary: Varies

INTERNSHIPS
How Many: Varies   **Duration:** Minimum of 1 month
Remuneration: Unpaid
Duties:
Clerical, research, and school talks. Duties mainly depend on intern interest and skills.

Qualifications:
Very smart, dedicated, and financially independent.

How to apply: Cover letter and resume

Ecological Society of America

Address: 2010 Massachusetts Avenue, NW, Suite 400
City: Washington   **State:** DC   **Zip:** 20036

Telephone: (202) 833-8773   **Fax:** (202) 833-8775
E-mail: esahq@esa.org
URL: http://www.sdsc.edu/SDSC/Research/Comp_Bio/ESA/ESA.html

Other locations: Ithaca, NY

Founded: 1915

Mission:
The Ecological Society of America promotes the science of ecology by facilitating the exchange of ideas among those interested in ecology and increasing public knowledge and understanding of the environment.

Programs:
Ecology, Ecological Applications, and *Ecological Monographs* are widely read and highly regarded research journals. The Society's *Bulletin* and newsletter keep members informed about matters of interest. The Society also sponsors an annual meeting that is attended by more than 2,000 individuals.

Atmosphere: Relaxed

Volunteers: 0 to 2

STAFF POSITIONS
Full-time staff: 20   **Part-time staff:** 3 to 6
Staff positions available: 0 to 1 per year
Staff salary: $18,000 per year plus benefits to start

INTERNSHIPS
How Many: 1 to 4   **Duration:** Usually 2 to 3 months
Remuneration: $1,500 to $2,000 total

Duties:
Specific projects, general office support, and participation in group efforts.

Qualifications:
Some university-level education in biology or ecology, or business administration or accounting.

How to apply: Cover letter and resume

Educational Foundation for Nuclear Science

Address: 6042 S. Kimbark Avenue
City: Chicago   **State:** IL   **Zip:** 60637

Telephone: (312) 702-2555   **Fax:** (312) 702-0725
E-mail: bullatomsci@igc.apc.org
URL: http://neog.com/atomic/

Other locations: None listed

Founded: 1945

Mission:
The Foundation's mission is to publish *The Bulletin of Atomic Scientists,* a bimonthly magazine focusing on nuclear security and international affairs. The keeper of the Doomsday Clock, the *Bulletin* is as likely to examine the role of the United Nations in peacekeeping as it is to dissect a nuclear arms control treaty.

Programs:
Besides the *Bulletin,* the Foundation also runs the Visiting Fellows Program. The program brings early-to-mid career journalists from nations with budding nuclear capabilities to the *Bulletin* offices.

Atmosphere: Casual, friendly, work oriented

Volunteers: 0

STAFF POSITIONS
Full-time staff: 9   **Part-time staff:** 1
Staff positions available: 0 to 1 per year
Staff salary: Depends on experience

INTERNSHIPS
How Many: 0   **Duration:** Not listed
Remuneration: Not listed
Duties:
Not listed.
Qualifications:
Not listed.

How to apply: Not listed

Educators for Social Responsibility

Address: 23 Garden Street
City: Cambridge   **State:** MA   **Zip:** 02138

Telephone: (617) 492-1764   **Fax:** (617) 864-5164
E-mail: natlesr@ix.netcom.com
URL: Not listed

Other locations: New York City and regional chapters

Founded: 1982

Mission:
Educators for Social Responsibility works to help young people develop the convictions and skills to shape a safe, sustainable, and just world.

Programs:
ESR is nationally recognized for promoting children's ethical and social development through its leadership in conflict resolution, violence prevention, intergroup relations, and character education. ESR supports educators and parents with professional development networks and instructional materials, and offers a wide range of training, resources, and consultation for preschool through adult settings. ESR's experiential and highly interactive programs are designed for teachers, counselors, administrators, parents, and community members. ESR produces a range of resources and educational opportunities.

Atmosphere: Fast-paced, supportive, friendly, democratically run, casual

Volunteers: Varies

STAFF POSITIONS
Full-time staff: 15   **Part-time staff:** 6
Staff positions available: Varies

Staff salary: Varies

INTERNSHIPS
How Many: Varies   **Duration:** Varies
Remuneration: Work-study positions are paid
Duties:
Varies depending on individual interest.

Qualifications:
ESR looks primarily for those interested in the work of the organization, as well as those who are hardworking, responsible, and have experience in the field of education and/or conflict resolution.

How to apply: Cover letter and resume; phone calls welcome

**JOBS
YOU CAN
LIVE WITH
1996**

Energy and Environmental Concepts, Inc.

Address: 325 S. Spruce Street
City: Traverse City   **State:** MI   **Zip:** 49684

Telephone: (800) 968-9998   **Fax:** (616) 947-4632
E-mail: Not listed
URL: Not listed

Other locations: None listed

Founded: 1980

Mission:
Energy and Environmental Concepts promotes stewardship of the earth, consistently striving to better educate children and to promote energy conservation, renewable resources, and environmentally safe products.

Programs:
Energy and Environmental Concepts programs revolve around: energy and environmental consulting, educational programs, environmental health, energy efficient lighting, color corrected and full spectrum lighting, light therapy equipment, commercial lighting and architectural design, and solar equipment.

Atmosphere: Relaxed

Volunteers: 2

STAFF POSITIONS
Full-time staff: 1   **Part-time staff:** 2
Staff positions available: 1 per year
Staff salary: Varies

INTERNSHIPS
How Many: 1 to 2   **Duration:** Varies
Remuneration: Varies
Duties:
Depends on specific project.
Qualifications:
An interest in related issues.

How to apply: Cover letter and resume

Environmental Action Foundation

Address: 6930 Carroll Avenue, Suite 600
City: Takoma Park   **State:** MD   **Zip:** 20912

Telephone: (301) 891-1100   **Fax:** (301) 891-2218
E-mail: eaf@igc.apc.org
URL: Not listed

Other locations: None listed

Founded: 1970

Mission:
The Foundation seeks to protect the environment through public education; to support the work of grassroots activists; and to bring our society in greater harmony with ecological and social justice imperatives through policy analysis, information education, technical assistance, coalition building, organizing, and litigation.

Programs:
Activities revolve around the following issues: environmental security, pollution, utilities, mass transit, corporate secrecy, climate change, global warming, toxics, reusable energy, and big money in politics.

Atmosphere: Casual and relaxed

Volunteers: 5

STAFF POSITIONS
Full-time staff: 15   **Part-time staff:** 2
Staff positions available: Varies
Staff salary: Varies

INTERNSHIPS
How Many: Varies   **Duration:** Varies
Remuneration: Unpaid
Duties:
Interns assist in the production of publications, work on specific environmental issues, and research information on selected topics.
Qualifications:
Initiative, the ability to work independently, and research and writing skills.

How to apply: Cover letter, resume, and application

Environmental Advantage

Address: 80 Wall Street, Suite 715
City: New York   **State:** NY   **Zip:** 10005

Telephone: (212) 482-0671   **Fax:** (212) 482-0679
E-mail: Not listed
URL: Not listed

Other locations: None listed

Founded: 1994

Mission:
The mission of Environmental Advantage is to develop businesses in the renewable energy and sustainable forest product industries. Efforts include organizing the demand for sustainable forest products and renewable energy systems and bringing together players in the financial and product value chains of renewable energy-sustainable forest product industries to increase product and capital flows.

Programs:
See mission statement.

Atmosphere: Informal and hectic

Volunteers: 0

STAFF POSITIONS
Full-time staff: 6   **Part-time staff:** 0
Staff positions available: 1 to 2 per year
Staff salary: Varies

INTERNSHIPS
How Many: Varies   **Duration:** 4 to 6 months
Remuneration: Varies
Duties:
A combination of research for projects and administrative support.

Qualifications:
An interest in business and the environment, oral and written communication skills, and relevant experience is helpful.

How to apply: Cover letter and resume

Environmental Law Institute

Address: 1616 P Street, NW, Suite 200
City: Washington **State:** DC **Zip:** 20036

Telephone: (202) 328-5150 **Fax:** (202) 328-5002
E-mail: Not listed
URL: Not listed

Other locations: None listed

Founded: Not listed

Mission:
For 25 years the Environmental Law Institute has played a pivotal role in shaping the fields of environmental policy, law, and management, both domestically and abroad. Today, ELI is an internationally recognized, independent research and education center.

Programs:
Through its information services, training courses and seminars, research programs, and policy recommendations, the Institute activates a broad constituency of environmental professions in government, industry, the private bar, public interest groups and academia. Central to ELI's mission is convening this diverse constituency to work cooperatively in developing effective solutions to pressing environmental problems.

Atmosphere: Casual, relaxed, hectic at times

Volunteers: 4

STAFF POSITIONS
Full-time staff: 60 **Part-time staff:** 0
Staff positions available: Varies
Staff salary: Varies

INTERNSHIPS
How Many: 3 to 5 **Duration:** 2 months to 1 year
Remuneration: Unpaid
Duties:
Research, administrative and clerical tasks.
Qualifications:
Writing and research skills, enthusiasm, and commitment.

How to apply: Cover letter, resume, academic transcript, and writing sample

Environmental Working Group

Address: 1718 Connecticut Avenue, NW, Suite 600
City: Washington   **State:** DC   **Zip:** 20009

Telephone: (202) 667-6982   **Fax:** (202) 232-2592
E-mail: info@ewg.org
URL: http://www.ewg.org

Other locations: None listed

Founded: 1993

Mission:
Environmental Working Group is a progressive, nonprofit, computer powered environmental organization that gives concerned citizens the information they need to protect their homes, communities, and planet Earth.

Programs:
The Environmental Working Group releases reports concerning current environmental issues every three to four weeks. Their Web page also is constantly updated for the most recent information.

Atmosphere: Casual, hectic, and dressier than most environmental organizations
Volunteers: 0

STAFF POSITIONS
Full-time staff: 14   **Part-time staff:** 1
Staff positions available: 1 to 2 per year
Staff salary: $18,000 to $27,000 per year

INTERNSHIPS
How Many: Unlimited   **Duration:** Varies
Remuneration: Unpaid
Duties:
Research, computer programming, attending meetings, and participating in every level of publication production.
Qualifications:
A strong interest in environmental issues.
How to apply: Cover letter, resume, and writing sample

Ethics and Public Policy Center

Address: 1015 15th Street, NW, Suite 900
City: Washington **State:** DC **Zip:** 20005

Telephone: (202) 682-1200 **Fax:** (202) 408-0632
E-mail: ethics@eppc.org
URL: Not listed

Other locations: None listed

Founded: 1976

Mission:
The Center was established to clarify and reinforce bonds between Judeo-Christian moral tradition and public debate over domestic and foreign policy issues. It affirms the political relevance of Western ethical imperatives in respect for the dignity of every person, individual freedom and responsibility, justice, the rule of law, and limited government. It maintains that moral reasoning is an essential complement to empirical calculation in the shaping of public policy.

Programs:
The Center has six primary areas of research and publications: religion and society, American citizenship and American identity, foreign policy, law and society, business, and culture and education. Three books published by the Center are: *Creation at Risk? Religion, Science, and Environmentalism*; *The Virgin and the Dynamo*; and *Death Without Dignity: How to Think About Euthanasia*.

Atmosphere: Formal

Volunteers: 0

STAFF POSITIONS
Full-time staff: 13 **Part-time staff:** 1
Staff positions available: Varies
Staff salary: Varies

INTERNSHIPS
How Many: 1 to 2 per semester **Duration:** Semester
Remuneration: Commuting expenses
Duties:
Research; analyzing, typing, and organizing research; creating status reports; attending and summarizing briefings; attending meetings; and contributing to manuscripts and articles.

Qualifications:
Senior undergraduate or graduate students with an understanding of our work and good writing and research skills.

How to apply: Cover letter, resume, and writing sample; phone calls welcome

Farm Sanctuary

Address: PO Box 150
City: Watkins Glen **State:** NY **Zip:** 14891

Telephone: (607) 583-2225 **Fax:** (607) 583-2041
E-mail: Not listed
URL: Not listed

Other locations: Oakland, CA

Founded: 1986

Mission:
Farm Sanctuary works to prevent animal abuse through hands-on rescue, investigations, exposures, public education programs, and legislation and litigation.

Programs:
The Farm Sanctuary Campaign Programs include: the "No Downer" campaign, which bans the marketing of "downed" animals (animals too sick to stand), public visitor programs and farm tours, a quarterly newsletter, and various educational videos.

Atmosphere: Busy and informal

Volunteers: Varies

STAFF POSITIONS
Full-time staff: 12 **Part-time staff:** 2
Staff positions available: 1 per year
Staff salary: $14,000 to $16,000 per year

INTERNSHIPS
How Many: 10 **Duration:** 1 to 3 months
Remuneration: Housing provided
Duties:
Duties vary from barn cleaning to bulk mailings and administrative duties to conducting educational tours.

Qualifications:
Dedication to animals and a commitment to vegetarianism.

How to apply: Cover letter, resume, and application

JOBS YOU CAN LIVE WITH 1996

Federation of American Scientists

Address: 307 Massachusetts Avenue, NE
City: Washington **State:** DC **Zip:** 20002

Telephone: (202) 546-3300 **Fax:** (202) 675-1010
E-mail: fas@fas.org
URL: http://www.fas.org/pub/gen/fas/

Other locations: None listed

Founded: 1945

Mission:
The Federation of American Scientists is a public-interest group that has worked to control the arms race, protect the environment, ensure the rights of scientists, promote international scientific exchange and, in general, prevent the misuse of science and technology in our society.

Programs:
Current publication projects in the Capitol Hill office include:
Arms Sales and Monitoring—a bimonthly newsletter.
Secrecy and Government Bulletin—a newsletter.
Space Policy.
PROMED—program for monitoring emerging diseases
Agriculture—a newsletter on micronutrients and one on the long-term global food situation
Truth and Power—a newsletter featuring science advice to governments.

Atmosphere: Varies

Volunteers: 0

STAFF POSITIONS
Full-time staff: 12 **Part-time staff:** 0
Staff positions available: Varies
Staff salary: Varies

INTERNSHIPS
How Many: Varies **Duration:** Varies
Remuneration: Unpaid
Duties:
There is no formal internship program, but they employ interns periodically as staff and project needs dictate.

Qualifications:
Varies.

How to apply: Cover letter and resume

Food and Water, Inc.

Address: RRI Box 68D
City: Walden **State:** VT **Zip:** 05873

Telephone: (802) 563-3300 **Fax:** (802) 563-3310
E-mail: Not listed
URL: Not listed

Other locations: None listed

Founded: 1986

Mission:
Food and Water, Inc. strives to educate and activate citizens on issues of food safety in order to eliminate unnecessary threats to the safety of the food supply and, ultimately, our health.

Programs:
Food and Water's campaigns focus on stopping food irradiation, use of toxic pesticides, and production of bio-engineered food products. They also publish a quarterly journal titled *Food and Water Journal*.

Atmosphere: Small, casual, hectic, pleasant

Volunteers: Varies

STAFF POSITIONS
Full-time staff: 4 **Part-time staff:** 3
Staff positions available: Varies
Staff salary: Varies, good benefits

INTERNSHIPS
How Many: Varies **Duration:** Varies
Remuneration: Unpaid
Duties:
Office work and phone networking.

Qualifications:
Interest in grassroots organizing and enthusiasm are a must; experience in related areas preferred.

How to apply: Cover letter, resume, and writing sample

Friends Committee on National Legislation (FCNL)

Address: 245 Second Street, NE
City: Washington   **State:** DC   **Zip:** 20002-5795

Telephone: (202) 547-6000   **Fax:** (202) 547-6019
E-mail: fcnl@igc.apc.org
URL: http://www.fcnl.org/pub/fcnl

Other locations: None listed

Founded: 1943

Mission:
FCNL brings the concerns, experiences, and testimonies of the Religious Society of Friends (Quakers) to policy decisions. FCNL advocates reconciliation among nations and peoples and works to eliminate militarism, coercion, and injustice. Efforts also promote civil rights, self-determination of Native Americans, restoration of confidence in government, economic and employment opportunities, and adequate housing, education, and health care.

Programs:
Current programs include:
EPI Centers—study groups on public policy issues.
Roots on Radio—helps people learn to promote progressive ideas on talk radio.
Major publications include: *FCNL Washington Newsletter* and *Indian Report*.

Atmosphere: Hectic, semicasual to professional dress

Volunteers: 6 to 8

STAFF POSITIONS
Full-time staff: 15   **Part-time staff:** 1
Staff positions available: Varies
Staff salary: $20,000 to $30,000 per year

INTERNSHIPS
How Many: 3   **Duration:** 11 months
Remuneration: $900 per month
Duties:
Tracking legislation, tracking political strategies for passing or blocking legislation, writing newsletter articles, and researching topics for senior staff.
Qualifications:
Good writing skills, good people skills, analytical thinking, and ability to learn quickly.

How to apply: Cover letter and resume; phone calls welcome.

Friends of the Boundary Water Wilderness

Address: 1313 SE Fifth Street, Suite 329
City: Minneapolis   **State:** MN   **Zip:** 55406

Telephone: (612) 379-3835   **Fax:** (612) 379-3842
E-mail: Not listed
URL: Not listed

Other locations: None listed

Founded: 1976

Mission:
FBWW focuses on the protection of the million-acre Boundary Water Canoe Area (BWCA) wilderness in northeastern Minnesota and the international Quetico-Superior Ecosystem in which it lies. The Friends use advocacy, analysis, and public education to protect the region's wilderness character, and help to promote proper management of the region on both sides of the international border to preserve the wilderness legacy for future generations.

Programs:
The Friends of the Boundary Waters produces the *BWCA Wilderness News*, which is a thrice yearly newsletter about environmental issues in the region. There is also a series of natural history brochures from around the area and a minimum-impact camping publication.

Atmosphere: Very casual, relaxed, can be busy

Volunteers: 2

STAFF POSITIONS
Full-time staff: 3   **Part-time staff:** 0
Staff positions available: None
Staff salary: Not listed

INTERNSHIPS
How Many: Varies   **Duration:** Varies
Remuneration: Unpaid
Duties:
Depends on specific project.
Qualifications:
Varies with specific project.
How to apply: Phone call

JOBS YOU CAN LIVE WITH 1996

Friends of the Earth

Address: 1025 Vermont Avenue, NW, Suite 300
City: Washington **State:** DC **Zip:** 20005

Telephone: (202) 783-7400 **Fax:** (202) 783-0444
E-mail: Foedc@igc.apc.org
URL: http://www.essential.org/Foe.html

Other locations: Seattle, WA and in 52 countries around the world

Founded: 1969

Mission:
Friends of the Earth is dedicated to protecting the planet from environmental degradation; preserving biological, cultural, and ethnic diversity; and empowering citizens to have an influential voice in decisions affecting the quality of their environment and their lives.

Programs:
Eco-Team: economics for the Earth—green taxes, federal budget, consumption, corporate accountability, and jobs.
Global Team: the international connection—trade and the environment, International Monetary Fund and World Bank reform.
Protect the Planet Team: pollution prevention and health—water protection, ozone protection, Northwest rivers and wetlands.
Journalism, Communications, and Marketing Team: getting the word out—*Friends of the Earth* newsmagazine, media, membership, events.

Atmosphere: Casual dress, dedicated people, relaxed to hectic

Volunteers: 4 to 5

STAFF POSITIONS
Full-time staff: 21 **Part-time staff:** 3
Staff positions available: 1 to 3 per year
Staff salary: $18,000 per year plus benefits

INTERNSHIPS
How Many: Varies **Duration:** Varies
Remuneration: Varies
Duties:
Work closely with project directors to research, write, lobby, and assist with administrative support.

Qualifications:
Relevant academic experience, commitment to promoting cultural diversity in environmental justice, excellent speaking and writing skills, computer literacy (including Word Perfect).

How to apply: Resume, cover letter, three references, and a writing sample

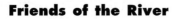

Friends of the River

Address: 128 J Street, 2nd Floor
City: Sacramento   **State:** CA   **Zip:** 95814

Telephone: (916) 442-3155   **Fax:** (916) 442-3396
E-mail: ftr@igc.apc.org
URL: Not listed

Other locations: None listed

Founded: 1970

Mission:
Friends of the River identifies outstanding rivers in California and mobilizes positive citizen action to save them. Friends of the River works in collaboration with individuals, grassroots organizations and local, state, and federal agencies to preserve, protect and restore the ecosystems of these rivers as they flow within or through California.

Programs:
Friends of the River programs involve protecting and restoring river ecosystems throughout California, organizing wild and scenic designation campaigns, being an information resource, advocating for sound state and federal water policies, and publishing a quarterly newsletter to members called *Headwaters*.

Atmosphere: Casual, small, friendly, often hectic

Volunteers: 450

STAFF POSITIONS
Full-time staff: 10   **Part-time staff:** 3 to 6
Staff positions available: 0 to 2 per year
Staff salary: $18,000 to $25,000 per year, health/dental insurance

INTERNSHIPS
How Many: 2 during school year   **Duration:** 3 to 9 months
Remuneration: $7 per hour to start, summer negotiable
Duties:
Administrative assistance, data entry, volunteer and other conservation project coordination.

Qualifications:
Ability to work independently, strong writing and speaking skills, and an interest in the environmental field.

How to apply: Cover letter, resume, and writing sample

Fund for Animals, Inc.

Address: 850 Sligo Avenue, Suite 300
City: Silver Spring **State:** MD **Zip:** 20910

Telephone: (301) 585-2591 **Fax:** (301) 585-2595
E-mail: fund4animals@igc.apc.org
URL: http://environlink.org/arrs/fund/

Other locations: New York, NY; Murchison, TX; Royal Oak, MI; Ann Arbor, MI; Houston, TX; San Francisco, CA; Albany, NY; and Louisville, KY
Founded: 1967

Mission:
Although the Fund for the Animals works on a variety of animal protection issues, the primary focus of their organization is the abolition of sport hunting and the mitigation of dog and cat overpopulation.

Programs:
Current campaigns focus on ending the trophy hunting of tundra swans, the orphaning of baby squirrels, the spring bail and hound hunting of black bears, Pennsylvania's live pigeon shoots, Alaska's wolf slaughter, and other egregious forms of "sport" hunting. The Fund also directs the Have-a-Heart program—one of the largest spay/neuter assistance programs. The Fund's publications include newsletters, updates and alerts, and their publication *Wildlife Watch*.

Atmosphere: Casual, hectic at times

Volunteers: Varies

STAFF POSITIONS
Full-time staff: 9 **Part-time staff:** 1
Staff positions available: Very few per year
Staff salary: Varies

INTERNSHIPS
How Many: Varies **Duration:** 3 months or 1 year
Remuneration: $500 or non-paid college credit
Duties:
Depending on needs and skills, the intern may be asked to execute a variety of tasks in his/her given area. These include legal and wildlife research, legislative work, and assisting with campaigns.

Qualifications:
Enthusiastic about working on animal-related issues, self-starter, and able to adapt to the atmosphere of the office.

How to apply: Cover letter, resume, application, writing sample, interview, and two recommendations

Global Change

Address: 1347 Massachusetts Avenue, SE
City: Washington   **State:** DC   **Zip:** 20003-1540

Telephone: (202) 547-0850   **Fax:** (202) 547-0850
E-mail: nsundt@igc.apc.org.
URL: http://solstice.crest.org/environment/global.change/gc.htm

Other locations: None listed

Founded: 1995

Mission:
Global Change works to produce and market a magazine and on-line publication dealing with global changes, climate change, and ozone depletion.

Programs:
All programs revolve around the production of the hardcopy quarterly version of *Global Change* magazine and also the development of an electronic version.

Atmosphere: Casual

Volunteers: 0

STAFF POSITIONS
Full-time staff: 1   **Part-time staff:** 0
Staff positions available: None
Staff salary: Varies

INTERNSHIPS
How Many: Not listed   **Duration:** Not listed
Remuneration: Not listed
Duties:
Assortment of work from filing to research and writing.

Qualifications:
Superior research and writing skills with an interest in specific issues.

How to apply: Cover letter, resume, and writing sample; phone calls welcome

Global Exchange

Address: 2017 Mission Street, #303
City: San Francisco   **State:** CA   **Zip:** 94110

Telephone: (415) 255-7296   **Fax:** (415) 255-7498
E-mail: globalexch@igc.org
URL: Not listed

Other locations: Berkeley, CA and Mill Valley, CA

Founded: 1988

Mission:
Global Exchange works to create more justice and economic opportunity in the world. The heart of the organization's work is the involvement of thousands of supporters around the country.

Programs:
Reality Tours—provides tours of countries such as Haiti, South Africa, and Brazil; includes interaction with the native citizens.
Public Education—publishes books and pamphlets on world hunger, trade, World Bank, and other important issues.
Fair Trade—helps build economic justice by promoting alternative trade that benefits low-income producers and artisan co-ops.
Material Assistance—provides money and technical support to successful grassroots groups in poor countries.
Human Rights Work—election observation teams, human rights reports, and brings long-term volunteers into conflict zones.

Atmosphere: Casual and busy

Volunteers: 10

STAFF POSITIONS
Full-time staff: 20   **Part-time staff:** 5
Staff positions available: 2 per year
Staff salary: Varies

INTERNSHIPS
How Many: Varies   **Duration:** Minimum of 2 months
Remuneration: Unpaid
Duties:
Varies according to specific program of interest.

Qualifications:
Self-motivated, creative and willing to make a set time commitment depending on the project.

How to apply: Cover letter, resume, application, and two letters of recommendation

Government Accountability Project

Address: 810 First Street, NE, Suite 630
City: Washington   **State:** DC   **Zip:** 20002

Telephone: (202) 408-0034   **Fax:** (202) 408-9855
E-mail: gap@igc.apc.org
URL: Not listed

Other locations: Seattle, WA

Founded: 1978

Mission:
The Government Accountability Project is a private, nonprofit, nonpartisan public interest group that provides legal representation for whistleblowers—corporate and government employees who speak out against fraud, waste, abuse of authority, environmental and public health and safety problems—and seeks resolution of the substantive concerns raised by whistleblowers.

Programs:
The Government Accountability Project: mounts legal campaigns to defend free speech rights of whistleblowers, providing legal representation; promotes successful resolution of the problems exposed by seeking forums for whistleblower's dissent; advocates for stronger legislative and judicial whistleblower protection through litigation and legislative education; and aids affected individuals and community groups in addressing issues raised by whistleblowing disclosures. Publications include *Courage Without Martyrdom* and the quarterly newsletter *Bridging the Gap*.

Atmosphere: Mostly relaxed and collegial

Volunteers: 15

STAFF POSITIONS
Full-time staff: 19   **Part-time staff:** 0
Staff positions available: Usually none
Staff salary: $22,000 to $26,000 per year to start

INTERNSHIPS
How Many: 5 to 15   **Duration:** Minimum of 10 weeks
Remuneration: Only graduate work-study
Duties:
Depends on specific projects.
Qualifications:
Commitment to public interest and ability to work independently.

How to apply: Cover letter, resume, and writing sample

Greater Yellowstone Coalition

Address: PO Box 1874, 13 S. Willson
City: Bozeman **State:** MT **Zip:** 59715

Telephone: (406) 586-1593 **Fax:** (406) 586-0851
E-mail: gyc@desktop.org
URL: http://www.idaho.net/gyc

Other locations: Dubois, WY; Cody, WY; and Idaho Falls, ID

Founded: 1983

Mission:
The Greater Yellowstone Coalition works to protect and preserve the greater Yellowstone ecosystem and the quality of life it sustains.

Programs:
Major programs revolve around: mining issues, geothermal energy, biodiversity, grizzlies, wolves, Endangered Species Act, private land issues, public lands and grazing issues, public lands and salvage logging, and forest planning. Publications include *Sustaining Greater Yellowstone—a Blueprint for the Future*, *Greater Yellowstone Report*, *Environmental Profile of the Greater Yellowstone Ecosystem*, and various reports and newsletters.

Atmosphere: Casual and hectic

Volunteers: 1

STAFF POSITIONS
Full-time staff: 23 **Part-time staff:** 3
Staff positions available: None
Staff salary: Varies

INTERNSHIPS
How Many: 3 **Duration:** 3 months
Remuneration: Minimum wage
Duties:
Assisting program directors with field work, research, and administrative work.
Qualifications:
A concern for the environment and a knowledge of Yellowstone is preferable.

How to apply: Cover letter, resume, and writing sample; phone calls welcome

JOBS YOU CAN LIVE WITH 1996

Green Corps

Address: 29 Temple Place
City: Boston   **State:** MA   **Zip:** 02111

Telephone: (617) 426-8506   **Fax:** (617) 292-8057
E-mail: Greencorps@aol.com
URL: Not listed

Other locations: In 20 cities nationwide

Founded: 1990

Mission:
Green Corps is a one-year training program designed to train the next generation of environmental leaders. As a grassroots organizing staff, Green Corps augments public education, research, and lobbying of state and national environmental organizations. Green Corps is also a field school to train individuals as environmental organizers. At the end of the year they help place graduates in positions in the environmental movement.

Programs:
Green Corps participants are immersed in a thorough training that includes intensive classroom work, one-on-one supervision and field training. The Green Corps "core curriculum" is a schedule of five field campaigns conducted in conjunction with leading environmental organizations. They work with national groups such as the Sierra Club, Rainforest Action Network, Alliance to Prevent Childhood Lead Poisoning, and many more.

Atmosphere: Relaxed, casual, and busy

Volunteers: Many

STAFF POSITIONS
Full-time staff: 20   **Part-time staff:** 1
Staff positions available: Varies
Staff salary: $15,000 to $16,000, health benefits, student loan repayment program

INTERNSHIPS
How Many: Unlimited   **Duration:** Varies
Remuneration: Academic credit, possible stipend

Duties:
Most interns take on projects relating to campaign development, grassroots organizing, and/or media and research. Duties could include leadership, volunteer coordination, public outreach, fundraising, event coordinating, writing for publications, etc.

Qualifications:
Interns must be able to work well with groups, be motivated and creative, and be concerned about the environment.

How to apply: Cover letter and resume

Greenlining Institute

Address: 785 Market Street, 3rd Floor
City: San Francisco   **State:** CA   **Zip:** 94103

Telephone: (415) 284-7200   **Fax:** (415) 284-7210
E-mail: Not listed
URL: Not listed

Other locations: None listed

Founded: 1994

Mission:
The Greenlining Institute empowers low-income, minority, and disabled communities through leadership training, economic development, and advocacy.

Programs:
Many of their programs revolve around economic development. Specifically, the Institute focuses on policy-level advocacy and negotiations regarding banking and CRA, insurance reform, small minority business development, and telecommunications and info-highway access.

Atmosphere: Casual, family atmosphere; relaxed, busy at times

Volunteers: 0

STAFF POSITIONS
Full-time staff: 6   **Part-time staff:** 7
Staff positions available: Varies
Staff salary: $18,000 to $25,000 per year

INTERNSHIPS
How Many: 3 to 6   **Duration:** Maximum 2 years
Remuneration: $10 hourly or $1,600 monthly
Duties:
Writing, research, and community organizing.

Qualifications:
Strong communication skills, commitment to communities, strong work ethic, and a professional attitude.

How to apply: Cover letter, resume, academic transcript, writing sample, and two references

Greenwire

Address: 3129 Mount Vernon Avenue
City: Alexandria   **State:** VA   **Zip:** 22305

Telephone: (703) 518-4600   **Fax:** (703) 518-8702
E-mail: greenwire@apn.com
URL: http://www.apn.com

Other locations: None listed

Founded: 1991

Mission:
Greenwire is an environmental news daily, covering the top news on the full range of issues at the state, national, and international levels. *Greenwire* is a news publication and does not engage in advocacy or research on policy, but there is no place to learn more or do more writing on current environmental issues and events.

Programs:
Greenwire—a 12-page daily that focuses on environmental issues.
Greenwire International—a six-page daily that has an international focus.
Utility News and Natural Gas News—a 10-page daily.

Atmosphere: Casual, hectic, very supportive, busy

Volunteers: 0

STAFF POSITIONS
Full-time staff: 8   **Part-time staff:** 0
Staff positions available: About 4 per year
Staff salary: $18,000 to mid-$20's per year depending on ability and experience

INTERNSHIPS
How Many: 1   **Duration:** 3 to 6 months
Remuneration: $1000 per month

Duties:
News gathering, writing, and some grunt work.

Qualifications:
Highest standards of writing abilities, leadership, ability to grow, and a knowledge of related issues.

How to apply: Cover letter and resume; faxes welcome, but no calls

Habitat for Humanity International

Address: 121 Habitat Street
City: Americus **State:** GA **Zip:** 31709-3498

Telephone: (912) 924-6935 **Fax:** (912) 924-0641
E-mail: Not listed
URL: Not listed

Other locations: 1200 affiliates in the US; 200 affiliates in 50 other nations

Founded: Not listed

Mission:
Habitat for Humanity is a movement of individuals and groups working in partnership to build houses with those who otherwise would be unable to afford decent shelter.

Programs:
Habitat partners volunteer their construction and administrative skills with the vision of eliminating poverty housing from the face of the earth. The organization is divided into departments: human resources, construction, finance, development, communication services, media relations, executive, affiliates, campus chapters, and child care.

Atmosphere: Busy, nonprofit atmosphere

Volunteers: Hundreds

STAFF POSITIONS
Full-time staff: Varies **Part-time staff:** Varies
Staff positions available: Varies
Staff salary: Varies

INTERNSHIPS
How Many: Numerous **Duration:** Usually 3 months
Remuneration: Housing, stipend, insurance
Duties:
Duties are dependent on specific programs.

Qualifications:
North American construction interns must be at least age 16; for international positions, at least 21. Internal internship qualifications vary.

How to apply: Cover letter and resume

JOBS YOU CAN LIVE WITH 1996

HawkWatch International

Address: PO Box 660
City: Salt Lake City **State:** UT **Zip:** 84110

Telephone: (801) 524-8511 **Fax:** (801) 524-8520
E-mail: Not listed
URL: Not listed

Other locations: Albuquerque, NM

Founded: 1986

Mission:
HawkWatch International works to monitor and promote the conservation of eagles, hawks, and other birds of prey.

Programs:
Environmental Education—producing various educational materials on environmental issues which effect the birds of prey.
Research—studying raptor migration and bird of prey behavior.
Publications—including *RaptorWatch* and *HawkFlash* newsletters.

Atmosphere: Casual, with a busy, nonprofit environment

Volunteers: 30

STAFF POSITIONS
Full-time staff: 5 **Part-time staff:** 2
Staff positions available: Varies

Staff salary: $15,000 to $20,000 per year

INTERNSHIPS
How Many: 3 to 4 **Duration:** 6 months
Remuneration: $350 per month and housing
Duties:
Environmental education using live raptors in classrooms for civic groups and other interested parties.

Qualifications:
Education in biology or related field, creative, flexible, and good people skills.

How to apply: Cover letter and resume

Heal the Bay

Address: 2701 Ocean Park Boulevard, Suite 150
City: Santa Monica   **State:** CA   **Zip:** 90405

Telephone: (310) 581-4188   **Fax:** (310) 581-4195/6
E-mail: htb@earthspirit.org
URL: Not listed

Other locations: None listed

Founded: 1985

Mission:
Heal the Bay is a nonprofit advocacy group working for a safe and healthy Santa Monica Bay. The organization uses research, education, community action, and policy programs to achieve this goal. Now in their second decade, Heal the Bay continues to fight to enhance and restore Southern California beaches and coastal waters for human and marine life.

Programs:
Heal the Bay has two volunteer-driven programs. The Gulter Patrol Program includes all of the city of Los Angeles. The Pt. Dume project is a restoration of a nature preserve in Malibu. Heal the Bay has two publications, a newsletter produced for members and one produced for volunteers.

Atmosphere: Casual, yet with high energy

Volunteers: 700

STAFF POSITIONS
Full-time staff: 11   **Part-time staff:** 4
Staff positions available: 1 per year
Staff salary: Mid-twenties per year to start

INTERNSHIPS
How Many: Varies   **Duration:** 4 months or semester
Remuneration: Unpaid
Duties:
Internships are offered in the programs, communications, policy and issues departments.

Qualifications:
A self-starter with ability to work independently.

How to apply: Cover letter and resume

JOBS YOU CAN LIVE WITH 1996

Herbert Scoville Jr. Peace Fellowship

Address: 110 Maryland Avenue, NE, Suite 211
City: Washington   **State:** DC   **Zip:** 20002

Telephone: (202) 546-0795   **Fax:** (202) 546-5142
E-mail: scoville@clw.org
URL: http://www.clw.org/pub/clw/scoville/scoville.html

Other locations: None listed

Founded: 1987

Mission:
The Scoville Peace Fellowship was established to provide college graduates who have an interest or experience in arms control with an entree into the nonprofit community and a Washington perspective on arms control, disarmament, and international security, as well as the budgetary and environmental issues related to these concerns.

Programs:
Each participating organization has its own programs and publications.

Atmosphere: Organizations have own atmospheres, but often informal and busy

Volunteers: 0

STAFF POSITIONS
Full-time staff: 0   **Part-time staff:** 1
Staff positions available: Not listed

Staff salary: Varies

INTERNSHIPS
How Many: 2 to 4 per semester   **Duration:** 4 to 6 months
Remuneration: $1400 per month, insurance

Duties:
Fellows work for a Washington, DC nonprofit organization conducting research, writing, arranging and attending conferences and seminars, tracking legislation, and otherwise immersing themselves in the Washington world of arms control.

Qualifications:
A bachelor's degree with an educational background or interest in arms control and disarmament.

How to apply: Cover letter, resume, academic transcript, writing sample, and two recommendations

JOBS YOU CAN LIVE WITH 1996

Hispanic Association of Colleges and Universities (HACU)

Address: 1367 Connecticut Avenue, NW, 2nd Floor
City: Washington **State:** DC **Zip:** 20036

Telephone: (202) 467-0893 **Fax:** (202) 496-9177
E-mail: HNIPHACU@ix.nctom.com
URL: Not listed

Other locations: Washington, DC and San Antonio, TX

Founded: 1986

Mission:
The Hispanic Association of Colleges and Universities seeks:
To promote the development of member colleges and universities.
To improve access to and the quality of post-secondary educational opportunities for Hispanic students.
To meet the needs of business, industry, and government through the development and sharing of resources, information, and expertise.

Programs:
HACU National Internship Program (HNIP)—a 10-week summer program that exposes talented college students to a challenging set of professional and educational experiences in the federal sector. It is a comprehensive program that provides a stipend and travel to and from the job site.
HACU also provides a number of educational programs through their San Antonio office, as well as a newsletter published on a monthly basis.

Atmosphere: Not listed

Volunteers: 0

STAFF POSITIONS
Full-time staff: 24 **Part-time staff:** 0
Staff positions available: Not listed
Staff salary: Not listed

INTERNSHIPS
How Many: 250 to 300 **Duration:** 10 weeks in summer
Remuneration: $370 to $490 per week

Duties:
Duties will vary according to the intern's academic background and the agency where they are placed.

Qualifications:
Applicants must be enrolled as either a graduate or undergraduate student and possess a minimum 3.0 GPA. Applicants are selected based on academic performance, community and campus activities.

How to apply: Resume, application, academic transcript, writing sample, and letter of recommendation from professor or advisor

Human Rights Watch (HRW)

Address: 1522 K Street, NW, #910
City: Washington **State:** DC **Zip:** 20005

Telephone: (202) 371-6592 **Fax:** (202) 371-0124
E-mail: HRWDC@HRW.ORG
URL: Not listed

Other locations: New York, NY; London, England; Brussels, Belgium; and Los Angeles, CA
Founded: 1978

Mission:
HRW conducts systematic, worldwide investigations of human rights abuses; defends freedom of thought and expression, due process, and equal protection of the law; documents and denounces murder, disappearances, torture, arbitrary imprisonment, exile, censorship, and other abuses of internationally recognized human rights. HRW has 5 regional divisions plus several project divisions.

Programs:
The projects are related to the mission statement, with the addition of The Arms Project, which attempts to document the link between weapons used in human rights abuses to the supply of those weapons through documentary field research. The Arms Project monitors and tries to prevent the transfer of small arms and military aid to governments that violate human rights and international law. Human Rights Watch also releases a number of publications on arms control issues and other related topics.

Atmosphere: Casual, fast-paced office

Volunteers: Varies

STAFF POSITIONS
Full-time staff: Over 100 **Part-time staff:** Varies
Staff positions available: Varies

Staff salary: $22,500 to $30,000 per year

INTERNSHIPS
How Many: Varies **Duration:** Semester or summer
Remuneration: Most are unpaid
Duties:
Intern duties include: administrative work, congressional hearings, research media for relevant stories; some interns conduct their own research projects.

Qualifications:
Interest in international affairs, human rights, good organizational and computer skills.

How to apply: Cover letter and resume; phone calls welcome

Information Data Management

Address: 9701 W. Higgins Road, Suite 500
City: Rosemont   **State:** IL   **Zip:** 60018

Telephone: (708) 825-2300   **Fax:** (708) 825-2303
E-mail: Not listed
URL: Not listed

Other locations: None listed

Founded: 1977

Mission:
Information Data Managements has a 3-point mission:
1. To develop and support quality software products for customers in medical and other related industries.
2. To create a positive work atmosphere where employees are valued and treated fairly, and where teamwork and mutual support are part of the normal environment.
3. To assure growth and profitability.

Programs:
Software Quality—committees focus on process improvements as they relate to training, project planning, and priority setting.
Joint Application Design (JAD)—process followed when creating new products for customers. During JAD sessions, IDM meets with customers to describe the functionality of a product and to get their feedback regarding what they are looking for.
Function Points—used to improve project estimation. This is a universal metric that measures the functionality of a product.
IDM also produces a customer newsletter titled *IDM Select*, which focuses on company, industry, and products/services updates.

Atmosphere: Casual, team-oriented, professional

Volunteers: 0

STAFF POSITIONS
Full-time staff: 95   **Part-time staff:** 3
Staff positions available: 12 to 15 per year
Staff salary: Based on experience, benefits included

INTERNSHIPS
How Many: Not listed   **Duration:** Not listed
Remuneration: Not listed
Duties:
Not listed.
Qualifications:
Not listed.
How to apply: Not listed

Innovative System Developers, Inc. (ISD)

Address: 5950 Symphony Woods Road, Suite 311
City: Columbia **State:** MD **Zip:** 21044

Telephone: (410) 997-3358 **Fax:** (410) 997-3580
E-mail: Not listed
URL: Not listed

Other locations: None listed

Founded: 1989

Mission:
ISD combines an understanding of customer needs with technical know-how to help organizations gain the capabilities and benefits offered by modern technology. ISD operates on the principle that the benefits of GIS technology can extend throughout an organization, takes exception to the conventional notion that one must be an expert in order to use the technology, and strives to deliver GIS functionality in a form accessible to a broad end-user audience.

Programs:
ISD provides GIS professional services and products to local governments, utilities, and private organizations. Projects include system development for a variety of applications: assessment and taxation, regional planning, election redistricting, demographic analysis, water/wastewater facility management, forestry, health care provision, school administration, geographic data translation, and data quality control.

Atmosphere: Casual, relaxed, and busy

Volunteers: 0

STAFF POSITIONS
Full-time staff: 7 **Part-time staff:** 1
Staff positions available: 2 to 3 per year
Staff salary: Varies depending on experience

INTERNSHIPS
How Many: 1 to 2 **Duration:** Not listed
Remuneration: Unpaid
Duties:
Depends on specific project and intern interest.

Qualifications:
Varies according to project.

How to apply: Cover letter and resume

Institute for Development Anthropology (IDA)

Address: 99 Collier Street, PO Box 2207
City: Binghamton **State:** NY **Zip:** 13902-2207

Telephone: (607) 772-6244 **Fax:** (607) 773-89930
E-mail: mhorowi@bingsuns.cc.binghamton.edu
URL: Not listed

Other locations: None listed

Founded: 1976

Mission:
IDA applies the comparative and holistic methodologies and theories of anthropology to improving the conditions of the world's poor through environmentally sustainable development, equitable economic growth, and respect for human rights and cultural pluralism.

Programs:
The Institute undertakes long-term field research as well as shorter-term focused missions in Africa, the Middle East, Asia, Eastern Europe, the newly independent republics of the former Soviet Union, Latin America, and the Caribbean, and among Native American populations. IDA publishes a monograph series in association with Westview Press; a Working Paper series; and *The Development Anthropologist*, a twice-yearly periodical.

Atmosphere: Chaotic, pressured, and friendly

Volunteers: Varies

STAFF POSITIONS
Full-time staff: 10   **Part-time staff:** 10
Staff positions available: Varies
Staff salary: Varies

INTERNSHIPS
How Many: Varies   **Duration:** Varies
Remuneration: Varies
Duties:
Duties depend on intern experience and qualifications. Interns participate in the full range of IDA's research and educational activities.
Qualifications:
Commitment to the goals of the Institute, enthusiasm, creativity, and language skills.

How to apply: Cover letter, resume, and writing sample

Institute for Global Communications (IGC)

Address: 18 De Boom Street
City: San Francisco   **State:** CA   **Zip:** 94107

Telephone: (415) 442-0220   **Fax:** (415) 546-1794
E-mail: info@igc.apc.org
URL: http://www.igc.apc.org/

Other locations: Washington, DC

Founded: 1986

Mission:
IGC seeks to serve, expand, and inspire movements of peace, economic and social justice, human rights, and environmental sustainability around the world by providing and developing accessible computer networking tools. IGC builds and supports computer systems serving nongovernmental, nonprofit organizations outside the US, particularly in regions not served by commercial communications companies.

Programs:
Providing Services—communication, information sharing, and electronic publishing services, including full access to the Internet.
Consulting—solves computer networking problems and expands the scope and depth of users' activities through the Internet.
Training—helps users take control of computer networking and information technology by learning effective and creative uses.
Adapting Technology—adapts computer technology and develops appropriate applications to solve new problems and to expand the possibilities of the computer networks.

Atmosphere: Casual, hectic

Volunteers: Several

STAFF POSITIONS
Full-time staff: 31   **Part-time staff:** 11
Staff positions available: 5 per year
Staff salary: $22,500 to $30,000 per year to start

INTERNSHIPS
How Many: Varies   **Duration:** Minimum 3 months
Remuneration: Discounted or free on-line time
Duties:
Building Web pages and related electronic publishing projects, on-line content coordination, representing IGC at events and conferences, training.

Qualifications:
Some familiarity with e-mail and Internet and an interest in environmental and progressive issues.

How to apply: Cover letter and resume; phone calls welcome

Institute for Global Ethics

Address: PO Box 563
City: Camden   **State:** ME   **Zip:** 04843

Telephone: (207) 236-6658   **Fax:** (207) 236-4014
E-mail: ethics@midcoast.com
URL: http://www.sourcemaine.com/ethics

Other locations: London, England and Auckland, New Zealand

Founded: 1990

Mission:
The Institute seeks to discover and articulate the global common ground of ethical values, analyze ethical trends, gather and disseminate information on global ethics, elevate public awareness and discussion of global ethical issues.

Programs:
Projects of the Institute include: a membership program, Ethical Fitness Seminar Program, an international program, character education and the Ethics in Education program, and publishing and publications. There is a periodical named *Insights on Global Ethics*, which is produced 10 times a year. Books, audio, and video tapes concerning similar issues are also available.

Atmosphere: Casual, very busy

Volunteers: Varies

STAFF POSITIONS
Full-time staff: 12   **Part-time staff:** 2
Staff positions available: Varies
Staff salary: Varies

INTERNSHIPS
How Many: Varies   **Duration:** 6 weeks
Remuneration: Varies, no stipend
Duties:
Special projects, writing, research, and general office support.
Qualifications:
Excellent communications skills; ability to work in busy office, both independently and as a team player.

How to apply: Cover letter and resume

Institute for International Cooperation and Development (IICD)

Address: PO Box 103
City: Williamstown   **State:** MA   **Zip:** 01267

Telephone: (413) 458-9828   **Fax:** (413) 458-3323
E-mail: IICD1@berkshire.net
URL: http://www.berkshire.net/~iicd1

Other locations: None listed

Founded: 1986

Mission:
The Institute for International Cooperation and Development is dedicated to promoting global understanding and international solidarity.

Programs:
IICD trains volunteers to work abroad in the following positions for 6 to 18 months, with preparation and follow-up periods in the US:
Mozambique—teach in rural schools for street children and families, and in vocational schools.
Zimbabwe—teach in a school for troubled youth and do construction work in a community.
Angola—plant trees, teach in primary and secondary schools, lead preschool activities, take part in community health projects.
Nicaragua and Brazil—community construction work and studies in other parts of the country.

Atmosphere: Varies

Volunteers: 1 to 2

STAFF POSITIONS
Full-time staff: 6   **Part-time staff:** 0
Staff positions available: 3 per year
Staff salary: Varies

INTERNSHIPS
How Many: Varies   **Duration:** 6 to 18 months
Remuneration: Unpaid
Duties:
Related to programs in the specific countries.
Qualifications:
Over 18 years old and ready to take on new challenges.

How to apply: Application; phone calls welcome

Institute for Local Self-Reliance

Address: 2425 18th Street, NW
City: Washington   **State:** DC   **Zip:** 20009-2096

Telephone: (202) 232-4108   **Fax:** (202) 332-0463
E-mail: ilsr@igc.apc.org
URL: Not listed

Other locations: Minneapolis, MN

Founded: 1974

Mission:
The Institute seeks to provide the conceptual framework, strategies, and information for the creation of ecologically sound and equitable communities. The mission is to lay the groundwork for developing humanly-scaled, sustainable economic systems.

Programs:
The Institute conducts research, undertakes policy analysis, and provides technical assistance to citizen groups, local government, and businesses in the areas of recycling, economic development, and the carbohydrate economy.

Atmosphere: Professional, busy, casual

Volunteers: Varies

STAFF POSITIONS
Full-time staff: 18   **Part-time staff:** 0
Staff positions available: 1 to 2 per year
Staff salary: $20,000 to $25,000 per year

INTERNSHIPS
How Many: Not listed   **Duration:** Varies
Remuneration: Unpaid
Duties:
Not listed

Qualifications:
Not listed

How to apply: Not listed

Institute for Mental Health Initiatives (IMHI)

Address: 4545 42nd Street, NW, Suite 311
City: Washington **State:** DC **Zip:** 20016

Telephone: (202) 364-7111 **Fax:** (202) 363-3891
E-mail: Not listed
URL: Not listed

Other locations: None listed

Founded: 1982

Mission:
A not-for-profit organization, the Institute for Mental Health Initiatives (IMHI) uses a public health approach to promote mental health and prevent emotional disorders. Information is collected from leading experts and current research is synthesized and presented to the public through training workshops, public service announcements, videos, publications, community seminars, and media consultations.

Programs:
Channeling Children's Anger—built on the idea that teenagers can learn to channel anger into problem solving and positive action.
Channeling Parents' Anger—aims to prevent child abuse by teaching parents skills for constructive anger management and by providing information about child development and child rearing.
Foster Resilience—focus is on how those who live and work with youth can give them the tools and skills they need to learn and grow.
Publications—*Dialogue: Insights into Human Emotions for Creative Professionals* and *How to Help Children Deal With Disasters*.

Atmosphere: Busy, relaxed, friendly

Volunteers: 5

STAFF POSITIONS
Full-time staff: 5   **Part-time staff:** 2
Staff positions available: Varies
Staff salary: Varies

INTERNSHIPS
How Many: 5 per year   **Duration:** At least 2 months
Remuneration: Unpaid
Duties:
Doing literature searches, reading, and summarizing psychology and social sciences articles for their publications. Duties may also include letter writing, general public relations, and other various project work.

Qualifications:
Organized, dependable, productive, and friendly.

How to apply: Cover letter, resume, academic transcript, and writing sample

Institute for Peace and Justice

Address: 4144 Lindell Boulevard, #408
City: St. Louis **State:** MO **Zip:** 63108

Telephone: (314) 533-4445 **Fax:** (314) 533-1017
E-mail: Not listed
URL: Not listed

Other locations: None listed

Founded: 1970

Mission:
The Institute for Peace and Justice is an independent, interfaith, not-for-profit center that was created as a response to the realities of war, racism, and global economic injustice. The Institute's mission is to create resources and to provide learning experiences in peace education and social justice for educators, schools, religious leaders, institutions, families, and individuals, both in St. Louis and internationally.

Programs:
The Institute's programs include: the Parenting for Peace and Justice Network (for parents and families), the Families Against Violence Program, Educating for Peace and Justice, *Families Caring* (and four other books for families), Celebrating Racial Diversity (and other manuals for teachers), Journey Into Compassion (books and retreats), and solidarity projects in Nicaragua.

Atmosphere: Casual, caring, cooperative, busy

Volunteers: 5

STAFF POSITIONS
Full-time staff: 5   **Part-time staff:** 2
Staff positions available: Varies
Staff salary: $20,000 to $25,000 per year with health insurance

INTERNSHIPS
How Many: Usually 1   **Duration:** 3 months to 2 years
Remuneration: Stipend for a 2-year internship
Duties:
Working with educational/curriculum areas, violence prevention work with families, and advocacy work on racial and/or economic justice issues.

Qualifications:
Self-starter, initiative, willingness to learn and participate in a more cooperative atmosphere, committed to peace and justice.

How to apply: Cover letter and resume

JOBS YOU CAN LIVE WITH 1996

Institute for Policy Studies

Address: 1601 Connecticut Avenue, NW
City: Washington   **State:** DC   **Zip:** 20009

Telephone: (202) 234-9382   **Fax:** (202) 387-7915
E-mail: ipscomm@igc.apc.org
URL: Not listed

Other locations: None listed

Founded: 1963

Mission:
The Institute for Policy Studies is an educational and research organization devoted to addressing the issues of social, political, and economic equity in the world. It is a center for progressive thought and action. Independent of government yet constantly engaged in the making and remaking of government policy, IPS serves as crucible, think tank, and conscience.

Programs:
Working Groups—examine the issues encompassing the crisis in democracy in the US and the world, global security, foreign policy, race, poverty, Third World women and development, human rights in Latin America, housing, and the new international economic order.
The Social Action Leadership School for Activists—an educational program that offers skills training and in-depth policy seminars to progressive organizations and individuals. The mission is to provide practical skills, a forum for discussion and development of progressive issues, and a unique networking opportunity to DC-area activists.

Atmosphere: Casual, busy

Volunteers: 0

STAFF POSITIONS
Full-time staff: 30   **Part-time staff:** 2
Staff positions available: Not listed
Staff salary: Varies

INTERNSHIPS
How Many: Varies   **Duration:** 3 to 6 months
Remuneration: Unpaid
Duties:
Research, marketing, and general support.
Qualifications:
Depends on specific programs.
How to apply: Call for application

Institute for Resource and Security Studies

Address: 27 Ellsworth Avenue
City: Cambridge **State:** MA **Zip:** 02139

Telephone: (617) 491-5177 **Fax:** (617) 491-6904
E-mail: irss@igc.apc.org
URL: Not listed

Other locations: None listed

Founded: 1984

Mission:
The Institute for Resource and Security Studies conducts public education programs and performs technical and policy analysis. The objective of the Institute is to promote peace and international security, efficient use of natural resources, and protection of the environment.

Programs:
Their programs relate directly to their mission statement.

Atmosphere: Informal, sometimes hectic

Volunteers: 2

STAFF POSITIONS
Full-time staff: 3 **Part-time staff:** 2
Staff positions available: Not listed
Staff salary: Not listed

INTERNSHIPS
How Many: 2 to 3 per year **Duration:** Varies
Remuneration: Varies
Duties:
Varies, depending on project.

Qualifications:
Not listed

How to apply: Cover letter and resume

JOBS YOU CAN LIVE WITH 1996

Institute for Science and International Security (ISIS)

Address: 236 Massachusetts Avenue, NE, #500
City: Washington   **State:** DC   **Zip:** 20002

Telephone: (202) 547-3633   **Fax:** (202) 547-3634
E-mail: 71263.346@CompuServe.com
URL: Not listed

Other locations: None listed

Founded: 1993

Mission:
ISIS is dedicated to providing the public, press, and policymakers with clear analysis of scientific and policy issues affecting national and international security, including the problems of war; regional and global arms races; the spread of nuclear weapons; and the environmental, health and safety hazards of nuclear weapons production. ISIS aims to counter misinformation and to encourage the free exchange of ideas concerning science and technology uses.

Programs:
The Nuclear Non-Proliferation Project—aims to apply technical and political expertise to stop the spread of nuclear weapons; seeks the creation of a sound technical foundation for efforts aimed at limiting and halting the production of nuclear weapons and nuclear explosive materials by all countries.
The Nuclear Weapons Production Project—focuses on finding ways to bring the Department of Energy's nuclear weapons production complex and research and testing facilities into line with new global realities, and to dismantle and dispose of surplus nuclear weapons and fissile materials.

Atmosphere: Informal, hectic

Volunteers: 0

STAFF POSITIONS
Full-time staff: 3   **Part-time staff:** 1
Staff positions available: Varies
Staff salary: Varies

INTERNSHIPS
How Many: 1 per term   **Duration:** 3 to 5 months
Remuneration: $200 per month
Duties:
Attend meetings/hearings in Congress, track relevant legislation, draft research memos/fact sheets, help set up workshops and conferences, route daily clips, update mailing lists, help distribute publications.

Qualifications:
An interest in related issues, working knowledge of Windows-based software is important. Internet experience is a plus.

How to apply: Cover letter, resume, and three references

Institute of World Affairs

Address: 1321 Pennsylvania Avenue, SE
City: Washington **State:** DC **Zip:** 20003

Telephone: (202) 544-4141 **Fax:** (202) 544-5115
E-mail: bjohnson@iwa.org
URL: Not listed

Other locations: Salisbury, CT

Founded: 1924

Mission:
The Institute provides diverse training and professional development programs through seminars, exchanges, residential conferences, and publications. Thousands of individuals from around the world have benefited from IWA's training programs, resulting in an international cadre of individuals dedicated to the Institute and its mission. The Institute promotes free exchange of ideas and does not adhere to any particular ideology or way of life.

Programs:
The Institute helps the international diplomatic community address challenges faced by a rapidly changing, post-Cold War world through regular training sessions for mid-level diplomats. The programs explore theories of conflict resolution, negotiation and mediation, and build negotiating skills through simulation exercises. The Institute conducts organizational and professional development programs for nongovernmental groups in Europe and the former Soviet Union interested in governance, civil society, finance, banking, environmental policy, and human health. The Institute publishes *IWA International*, a newsletter on international issues.

Atmosphere: Formal when necessary, otherwise casual

Volunteers: 4

STAFF POSITIONS
Full-time staff: 4 **Part-time staff:** 8
Staff positions available: 1 per year
Staff salary: $25,000 per year to start

INTERNSHIPS
How Many: 3 **Duration:** Varies
Remuneration: Academic credit is available
Duties:
General office duties and work on specific projects.

Qualifications:
Computer literacy, intelligence, and communication skills.

How to apply: Cover letter and resume

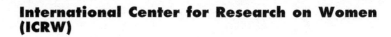

International Center for Research on Women (ICRW)

Address: 1717 Massachusetts Avenue, NW, Suite 302
City: Washington　　　　　**State:** DC　　**Zip:** 20037

Telephone: (202) 797-0007　　**Fax:** (202) 797-0200
E-mail: icrw@igc.apc.org
URL: Not listed

Other locations:　None listed

Founded: 1976

Mission:
ICRW undertakes policy research and advocacy activities in partnership with developing world institutions and international agencies. ICRW works for the improvement of women's economic and health status and toward women's participation in decisionmaking regarding the environment and governance.

Programs:
Policy research on issues of women and development in the Third World, especially women's economic empowerment, health, and the environment. Special topics include women, poverty and women-headed households. ICRW also publishes policy papers, reports-in-brief, and fact sheets on a number of related issues.

Atmosphere:　Professional, flexible

Volunteers: 0

STAFF POSITIONS
Full-time staff: 23　　　　　**Part-time staff:** 1
Staff positions available: 2 to 3 per year
Staff salary: $22,000 to $23,000 per year to start

INTERNSHIPS
How Many: Varies　　　　**Duration:** Varies
Remuneration: Unpaid
Duties:
Depends on intern interest and qualifications.

Qualifications:
Interest in related issues.

How to apply: Phone call

International Development and Energy Associates, Inc. (IDEA)

Address: 1111 14th Street, NW, Suite 710
City: Washington **State:** DC **Zip:** 20009

Telephone: (202) 289-4332 **Fax:** (202) 371-9015
E-mail: 76201.1772@compuserve.com
URL: Not listed

Other locations: Rosslyn, VA; Beltsville, MD; Moscow, Russia; Almaty, Kazakhstan; Kiev, Ukraine; and Yerevan, Armenia
Founded: 1981

Mission:
IDEA provides technical and analytical skills to energy sector entities including utilities, engineering firms, government agencies, and multilateral development organizations. IDEA's primary focus is on the developing countries of Asia, the former Soviet Union, and northern Africa.

Programs:
Programs are all related to IDEA's mission statement.

Atmosphere: Semiformal, sometimes relaxed

Volunteers: 0

STAFF POSITIONS
Full-time staff: 45 **Part-time staff:** 0
Staff positions available: 1 per year
Staff salary: Varies by job classification, experience, and education

INTERNSHIPS
How Many: Varies **Duration:** Not listed
Remuneration: Not listed
Duties:
Not listed
Qualifications:
Not listed

How to apply: Cover letter and resume

International Development Exchange (IDEX)

Address: 827 Valencia Street, #101
City: San Francisco **State:** CA **Zip:** 94110-1736

Telephone: (415) 824-8384 **Fax:** (415) 824-8387
E-mail: idex@igc.apc.org
URL: http://www.digimark.net/idex/overview.html

Other locations: None listed

Founded: 1985

Mission:
IDEX builds partnerships to overcome economic and social injustice, and works with people to bring greater control over the resources, political structures, and economic processes that affect their lives. IDEX identifies small-scale, grassroots development projects in Asia, Africa, and Latin America and links them with supportive US partners. IDEX also educates North Americans on the challenges and successes of marginalized people around the world.

Programs:
IDEX works on over 300 projects, focusing on 9 nations: Bangladesh, Ghana, Guatemala, India, Mexico, Mozambique, Nicaragua, Phillipines, and Zimbabwe. Typical projects are organized around microenterprise development, women's groups, agricultural development, or social service delivery. Their formal association with indigenous, nongovernmental organizations helps to maintain communication. The School Partnership Program brings international development experiences into 140 US classrooms, develops curricula to assist teachers in their instruction of global issues, and facilitates a link to local service learning projects.

Atmosphere: Casual, hard-working, sometimes crazy

Volunteers: 50

STAFF POSITIONS
Full-time staff: 6   **Part-time staff:** 1
Staff positions available: 1 per year
Staff salary: $20,000 to $24,000 per year, health/dental insurance

INTERNSHIPS
How Many: 12   **Duration:** 3 months
Remuneration: Unpaid
Duties:
Correspondence, database management, research, report writing, presentations, and more.
Qualifications:
Macintosh skills, interest and enthusiasm for international social justice issues.
How to apply: Phone call

International Institute for Energy Conservation (IIEC)

Address: 750 First Street, NE, Suite 940
City: Washington **State:** DC **Zip:** 20002

Telephone: (202) 842-3388 **Fax:** (202) 842-1565
E-mail: IIEC@IGC.APC.ORG
URL: Not listed

Other locations: London, England; Bangkok, Thailand; and Santiago de Chile

Founded: 1984

Mission:
IIED works to accelerate global adoption of energy efficiency policies, technologies, and practices to enable economic and environmentally sustainable development. IIED focuses on developing countries and the emerging nations of Central and Eastern Europe.

Programs:
IIEC's programs involve private sector initiatives, transportation, global energy efficiency initiative, climate change, multilateral development banks, and regional initiatives in Asia, Latin America, Europe, and South Africa. There is also a newsletter titled *E-Notes*, along with other various reports.

Atmosphere: Professional, casual

Volunteers: 3

STAFF POSITIONS
Full-time staff: 32 **Part-time staff:** 2
Staff positions available: 6 per year
Staff salary: Depends on experience

INTERNSHIPS
How Many: 2 to 4 **Duration:** Flexible
Remuneration: Varies
Duties:
Depends on specific area of interest.

Qualifications:
An interest in energy-related issues.

How to apply: Cover letter, resume, and writing sample

Issues in Science and Technology

Address: PO Box 830688, Mail Station AD 13
City: Richardson **State:** TX **Zip:** 75083

Telephone: (214) 883-6325 **Fax:** (214) 883-6327
E-mail: issuessn@utdallas.edu
URL: http://www.enews.com/

Other locations: Washington, DC

Founded: 1984

Mission:
Issues in Science and Technology is a journal that provides a forum where academic researchers, government officials, business leaders, and others with a stake in public policy can analyze and propose solutions to national problems that involve science, technology, and medicine. Experts have the opportunity to share their insights directly with a broad audience so that their knowledge can shape critical decisions.

Programs:
All programs revolve around the release of the publication.

Atmosphere: Academic atmosphere

Volunteers: 0

STAFF POSITIONS
Full-time staff: 4 **Part-time staff:** 1
Staff positions available: 0 to 1 per year
Staff salary: $19,000 per year at start

INTERNSHIPS
How Many: Not listed **Duration:** Not listed
Remuneration: Not listed
Duties:
Not listed

Qualifications:
Not listed

How to apply: Not listed

Izaak Walton League of America, Inc.

Address: 707 Conservation Lane
City: Gaithersburg   **State:** MD   **Zip:** 20878-2983

Telephone: (301) 548-0150   **Fax:** (301) 548-0149
E-mail: Cberger@iwla.org
URL: Not listed

Other locations: Minneapolis, MN

Founded: 1922

Mission:
The Izaak Walton League of America, Inc. is dedicated to the conservation of clean water, clean air, and soil conservation.

Programs:
Outdoor America—a quarterly magazine.
Outdoor Ethics—a program with quarterly magazines.
Save Our Streams—water quality monitoring program throughout the US.
Carrying Capacity—a sustainability program.
Energy Program—only in the Minneapolis office.

Atmosphere: Casual, relaxed, hectic at times

Volunteers: 2

STAFF POSITIONS
Full-time staff: 25   **Part-time staff:** 3
Staff positions available: 0 to 2 per year
Staff salary: Varies, all include full benefits

INTERNSHIPS
How Many: 1 to 2   **Duration:** 3 months
Remuneration: Varies
Duties:
Visiting and assisting at monitoring sites, media calls, computer work.

Qualifications:
Varies, but an interest in conservation is crucial.

How to apply: Cover letter, resume, and writing sample

JOBS YOU CAN LIVE WITH 1996

Just Harvest: A Center for Action Against Hunger

Address: 120 E. 9th Avenue
City: Homestead   **State:** PA   **Zip:** 15120

Telephone: (412) 464-0739   **Fax:** (412) 464-0758
E-mail: Not listed
URL: Not listed

Other locations: None listed

Founded: 1987

Mission:
Just Harvest is a nationwide membership organization that promotes economic justice and works to influence public policy and to educate, empower, and mobilize the citizens of their community towards the elimination of hunger.

Programs:
Just Harvest programs help mobilize grassroots lobbying, build coalitions; plan direct-action campaigns to expand school breakfast programs; empower welfare recipients; get state, city, and county governments to address hunger, oppose federal cuts, and expand summer food programs. Just Harvest also produces *Just Harvest News*, which discusses similar issues.

Atmosphere: Causal, relaxed, hectic

Volunteers: 3

STAFF POSITIONS
Full-time staff: 5   **Part-time staff:** 0
Staff positions available: None
Staff salary: $19,000 per year with annual raises

INTERNSHIPS
How Many: 3 to 5 per semester   **Duration:** Varies
Remuneration: Unpaid
Duties:
Clerical work and other specialized work on specific projects.
Qualifications:
An interest in related issues.

How to apply: Cover letter, resume, and writing sample; phone calls welcome

Land Institute

Address: 2440 E. Water Well Road
City: Salina   **State:** KS   **Zip:** 67401

Telephone: (913) 823-5376   **Fax:** (913) 823-8728
E-mail: theland@igc.apc.org
URL: Not listed

Other locations: Matfield Green, KS

Founded: 1976

Mission:
The Land Institute, a nonprofit research and educational organization, is devoted to sustainable agriculture and good stewardship of the earth. The Land Institute offers a unique post-graduate internship program, serves as a center for the study of environmental and agricultural issues, and conducts pioneering research into the development of sustainable agriculture and communities based on the model of the prairie.

Programs:
Research at the Land Institute envisions sustainable agriculture modeled on the prairie ecosystem, one less dependent on fossil fuels and chemicals, one more conserving of water and soil and of human communities. The Land Institute also publishes *The Land Report*, which features intern-written articles and illustrations on their research and is published three times a year.

Atmosphere: Relaxed

Volunteers: 2

STAFF POSITIONS
Full-time staff: 14   **Part-time staff:** 1
Staff positions available: 2 per year
Staff salary: $14,000 to $30,000 per year

INTERNSHIPS
How Many: 8   **Duration:** 10 months
Remuneration: $598 per month
Duties:
Each intern chooses an experiment that has been designed by staff as part of the on-going natural systems agriculture and Sunshine Farm research.
Qualifications:
College graduates or upper-level students, good health, stamina, and a love for working outside.

How to apply: Cover letter, resume, personal essay, transcripts, and two letters of recommendation

Land Stewardship Project (LSP)

Address: 2200 4th Street
City: White Bear Lake   **State:** MN   **Zip:** 55110

Telephone: (612) 653-0618   **Fax:** (612) 653-0589
E-mail: Not listed
URL: Not listed

Other locations: Lewiston, MN and Monterideo, MN

Founded: 1982

Mission:
LSP is devoted to fostering an ethic of stewardship toward farmland. Work focuses on developing and promoting sustainable communities and a system of agriculture that is environmentally sound, economically viable, and socially just.

Programs:
1000 Friends of Minnesota—land use and urban sprawl issues.
Factory Farm Awareness Campaign—organizing and educating about large-scale livestock containment operations and promoting sustainable alternatives.
Holistic Resource Management—seminars in process of long-term planning for farmers.
Farmers Accessing Credit Together—local citizens holding banks accountable to the Community Reinvestment Act.

Atmosphere: Casual, hectic

Volunteers: 20

STAFF POSITIONS
Full-time staff: 14   **Part-time staff:** 0
Staff positions available: Varies
Staff salary: $17,000 to $24,000 per year plus full benefits

INTERNSHIPS
How Many: Varies   **Duration:** Varies
Remuneration: Varies
Duties:
Depends on funding sources, interns often focus on community organizing.
Qualifications:
Smart, some experience or background in agriculture or organizing.

How to apply: Phone call

Lawyers Alliance for World Security

Address: 1601 Connecticut Avenue, NW, Suite 600
City: Washington **State:** DC **Zip:** 20009

Telephone: (202) 745-2450 **Fax:** (202) 667-0444
E-mail: Not listed
URL: Not listed

Other locations: None listed

Founded: 1981

Mission:
The Lawyers Alliance for World Security is a nonprofit, nonpartisan membership organization of legal professionals dedicated to stopping unrestrained weapons proliferation and bringing the rule of law to the newly independent nations of the former Soviet Union.

Programs:
Current programs include: advocacy studies on legal issues related to ending the proliferation of mass destruction and conventional weapons around the world; and Lawmaking for Democracy, a program aimed at bringing the rule of law to the former Soviet Union through the training of parliamentarians of the newly independent nations. In addition, the Alliance hosted a delegation from President Yeltsin's legal counsel's office in December 1993.

Atmosphere: Casual, sometimes hectic

Volunteers: 3

STAFF POSITIONS
Full-time staff: 5 **Part-time staff:** 0
Staff positions available: 0 to 1 per year
Staff salary: $21,000 per year

INTERNSHIPS
How Many: 2 to 3 **Duration:** 4 months
Remuneration: $100 or $500 per month
Duties:
Research on arms control and security issues; attending and reporting on Congressional hearings, briefings, and press conferences; preparing and coordinating mailings; filing; and handling communications.

Qualifications:
Quick learner, organizational skills, good writer, and relevant background.

How to apply: Cover letter, resume, academic transcript, writing sample, and two letters of recommendation

Lawyers Committee on Nuclear Policy

Address: 666 Broadway, #625
City: New York   **State:** NY   **Zip:** 10012

Telephone: (212) 674-7790   **Fax:** (212) 674-6149
E-mail: LCNP@aol.com
URL: Not listed

Other locations: The Hague, Netherlands

Founded: 1981

Mission:
The mission of the Lawyers Committee on Nuclear Policy is to use law to assist in disarmament, in particular nuclear disarmament.

Programs:
Major programs are: the World Court Case on legality of nuclear weapons (World Court Project), Limitation and Abolition of International Arms, a treaty for the prohibition and elimination of nuclear weapons, and the third Hague Peace Conference.

Atmosphere: Casual, busy, invigorating, friendly

Volunteers: Varies

STAFF POSITIONS
Full-time staff: 1   **Part-time staff:** 2
Staff positions available: 1 per year
Staff salary: $20,000 to $30,000 per year

INTERNSHIPS
How Many: 5 to 8   **Duration:** Flexible
Remuneration: Unpaid
Duties:
Research and policy development.

Qualifications:
Some knowledge of United Nations, disarmament issues, and international law. Also, a personal interest in peace.

How to apply: Cover letter, resume, and phone call

League of Conservation Voters (LCV)

Address: 1707 L Street, NW, Suite 750
City: Washington **State:** DC **Zip:** 20036

Telephone: (202) 785-8683 **Fax:** (202) 835-0491
E-mail: victoria_masotta@lcv.org
URL: http://www.lcv.org

Other locations: None listed

Founded: 1970

Mission:
The League of Conservation Voters is the bipartisan political arm of the environmental movement. They focus on electing pro-environmental candidates to Congress.

Programs:
LCV's trademark is the National Environmental Scorecard, which highlights the environmental voting record for every member of Congress per session. There are also programs that support pro-environment Congress members.

Atmosphere: Professional, relaxed, friendly

Volunteers: 0

STAFF POSITIONS
Full-time staff: 20 **Part-time staff:** 0
Staff positions available: 0 to 1 per year
Staff salary: Varies

INTERNSHIPS
How Many: 3 **Duration:** 6 months
Remuneration: Varies
Duties:
Approximately 70% programmatic work and 30% administrative.

Qualifications:
Interns must have initiative, be imaginative, possess a strong interest in electoral politics, Macintosh knowledge, word processing, spreadsheet applications, and Internet experience a plus.

How to apply: Cover letter, resume, and writing sample

Leveraged Outreach Project

Address: 1511 K Street, NW, Suite 1104
City: Washington **State:** DC **Zip:** 20005

Telephone: (202) 628-5922 **Fax:** (202) 628-5923
E-mail: sosdoby@access.digex.net
URL: Not listed

Other locations: None listed

Founded: 1995

Mission:
The Leveraged Outreach Project is an effort by several national religious, arms control, advocacy, and policy organizations that participate in the Arms Transfer Working Group. The Project's purpose is to increase public education about the implications of US arms export policies and to build public support for more conventional arms control.

Programs:
The Leveraged Outreach Project prepares materials for presentation to major nongovernmental organizations and arranges for speakers.

Atmosphere: Fast-paced, professional, and a good sense of humor

Volunteers: 0

STAFF POSITIONS
Full-time staff: 1   **Part-time staff:** 1
Staff positions available: None
Staff salary: None listed

INTERNSHIPS
How Many: 1   **Duration:** Varies
Remuneration: Possibly
Duties:
General office work, research, attendance at coalition meetings, and writing.

Qualifications:
Highly intelligent and mature, responsible, self-motivated, able to follow instructions, creative, sense of humor, and professional behavior.

How to apply: Cover letter, resume, academic transcript, and writing sample

Living Classrooms Foundation

Address: 717 Eastern Avenue, Pier 5 Lighthouse
City: Baltimore **State:** MD **Zip:** 21202

Telephone: (410) 685-0295 **Fax:** (410) 752-8433
E-mail: Not listed
URL: Not listed

Other locations: None listed

Founded: 1985

Mission:
Living Classrooms, a nonprofit, educational facility, provides hands-on, experience-based programs. The Foundation's philosophy is that students in small groups and environmentally challenging settings respond with a degree of learning rarely found in traditional classrooms. The educational and job-training programs are designed to motivate and empower youth to learn by doing, so that they may succeed academically in the work place and in their lives.

Programs:
The Foundation offers a variety of living classrooms, including the 104-foot pungy schooner, *Lady Maryland*; a 1948 historic buyboat *Mildred Belle*; the recently reconstructed skipjack, *Sigsbee*; the Maritime Institute job-training facility for at-risk youth; and a 100-acre llama farm. Programs, aimed primarily at middle school students, emphasize the history, economics, and ecology of the Chesapeake Bay region, as well as provide cultural enrichment, job training, and career development.

Atmosphere: The pace can be demanding for educators

Volunteers: Varies

STAFF POSITIONS
Full-time staff: 50 **Part-time staff:** 0
Staff positions available: None

Staff salary: Varies

INTERNSHIPS
How Many: Varies **Duration:** At least 3 months
Remuneration: Varies

Duties:
Interns work alongside experienced educational staff and are trained to lead a wide variety of hands-on activities. Focus is on interacting with, teaching, and supervising students of all ages and abilities.

Qualifications:
Enthusiastic, well-spoken, reliable, flexible, able to work with diverse groups and with minimal supervision, and interested in Chesapeake Bay ecology and history. Science or teaching background preferred.

How to apply: Cover letter, resume, and application

Loka Institute

Address: PO Box 355
City: Amherst **State:** MA **Zip:** 01004

Telephone: (413) 253-2828 **Fax:** (413) 253-4942
E-mail: LOKA@amherst.edu
URL: http://www.amherst.edu/~Loka

Other locations: None listed

Founded: 1987

Mission:
The Loka Institute seeks to make science and technology policy more attentive to social, political, and environmental concerns, and encourages the inclusion of grassroots and public-interest groups, workers, and citizens in the policy-making process.

Programs:
The Loka Institute established FASTnet (the Federation of Activists on Science and Technology Network). Loka also serves as the coordinator for a nationwide community research center, and is currently organizing a national campaign to democratize technology. The Loka Institute distributes *Loka Alerts*, which are essays on the democratic politics of science and technology. Other publications include *Democracy and Technology: Putting Science to Work for Communities*.

Atmosphere: Comfortable, casual, many discussions

Volunteers: 2

STAFF POSITIONS
Full-time staff: 2   **Part-time staff:** 1
Staff positions available: None usually
Staff salary: Varies

INTERNSHIPS
How Many: Varies   **Duration:** Varies
Remuneration: Stipend when funding permits
Duties:
Managing one of Loka's e-mail discussion lists, acting as "Web-master," conducting research on current and relevant social issues, fundraising, proposal writing, bookkeeping, general writing, and outreach.

Qualifications:
Computer literacy, ability to work on collaborative projects, self-motivation, flexibility, and creativity.

How to apply: Cover letter and resume; e-mail welcome

Los Niños

Address: 287 G Street
City: Chula Vista   **State:** CA   **Zip:** 91910

Telephone: (619) 426-9110   **Fax:** (619) 426-6664
E-mail: losninos@electriciti.com
URL: Not listed

Other locations: None listed

Founded: 1974

Mission:
Los Niños is dedicated to improving the quality of life for Mexican children and their families, and simultaneously, to provide education on the benefits of community self-development through cultural interaction.

Programs:
Nutrition/Health Program for families, Family Gardens Project to teach families how to grow fruits and vegetables in any space available, Forestation Program in the city of Tijuana, School Ecology/Gardening Project, and Development Education Program to teach US citizens about Mexican border relations and socioeconomic aspects of Mexican border cities.

Atmosphere: Casual, relaxed

Volunteers: 2

STAFF POSITIONS
Full-time staff: 10   **Part-time staff:** 40
Staff positions available: Varies
Staff salary: Varies

INTERNSHIPS
How Many: 12 summer, 2 longer   **Duration:** 6 weeks to 1 year
Remuneration: Unpaid
Duties:
Interns teach on related issues and assist staff with ongoing community development projects.

Qualifications:
Experience with teaching, ability to speak Spanish, interest in US-Mexico border relations, and cultural understanding.

How to apply: Application

Louisiana Nature Center

Address: 10601 Dwyer Road, PO Box 870610
City: New Orleans **State:** LA **Zip:** 70187-0610

Telephone: (504) 246-5672 **Fax:** (504) 242-1889
E-mail: Not listed
URL: Not listed

Other locations: New Orleans, LA

Founded: 1980

Mission:
The Center exhibits diversity of wildlife, preserves native Louisiana habitat, educates a diverse audience about the natural world, fosters a balanced approach to conservation, enhances care and survival of wildlife through research, provides opportunities for recreation in natural settings, operates a financially self-sufficient collection of facilities, and weaves entertainment value throughout the visitor experience.

Programs:
The hands-on, multi-disciplinary, programs are designed using the state's Department of Education K–6 Science Curriculum Guide Performance Objectives, and include: Preschool Environmental Education Programs (PEEP); Beyond the Window Sill (grades 1–2); Earth Lab Archeology, Wetland and Forest Ecology, and Wastewise (all target grades 3–9). Also offered are science overnights, scout merit badge programs, summer camps, public field trips, and teacher workshops and outreach programs. A school program guide, brochures for each specific program offered, trail guides, and other resources are available to the public and teachers.

Atmosphere: Casual, comfortable, professional

Volunteers: 52

STAFF POSITIONS
Full-time staff: 12 **Part-time staff:** 14
Staff positions available: Not listed

Staff salary: Not listed

INTERNSHIPS
How Many: 1 to 2 per semester **Duration:** 2 months
Remuneration: Unpaid

Duties:
Program and exhibit development, curriculum development, general naturalist duties including teaching assignments.

Qualifications:
Energetic, enthusiastic, dependable, responsible, majoring in education, biology, or similar field, good writing skills.

How to apply: Cover letter and resume; phone calls are welcome

Maine Organic Farmers & Gardeners Association (MOFGA)

Address: PO Box 2176
City: Augusta **State:** ME **Zip:** 04338

Telephone: (207) 622-3118 **Fax:** Not listed
E-mail: Not listed
URL: Not listed

Other locations: None listed

Founded: 1972

Mission:
The MOFGA mission is . . .
To grow safe and healthful food and make it available to all.
To protect natural resources on the farm and in the garden.
To promote sustainable farming and support rural communities.
To educate consumers about the connections among food, farming, and the environment.

Programs:
1. Educate farmers and gardeners by showing how they can protect and recycle natural resources.
2. Certify organic food, which tells the consumer that food has been grown according to strict standards that prohibit synthetic chemicals and require farming that enhances solid life, recycles plant and animal waste, and uses renewable resources.
3. Celebrate rural living by organizing the annual Common Ground Country Fair.
4. Publish the quarterly newspaper, *The Maine Organic Farmer and Gardener*, featuring a wealth of farm and garden information.

Atmosphere: Casual, relaxed

Volunteers: Over 1000

STAFF POSITIONS
Full-time staff: 6 **Part-time staff:** 0
Staff positions available: None

Staff salary: Varies

INTERNSHIPS
How Many: Up to 40 or more **Duration:** Varies
Remuneration: Depends on specific farmer
Duties:
Varies from farm to farm.

Qualifications:
None. This program is an educational program, and anyone can participate.

How to apply: Call for an application

Management Sciences for Health (MSH)

Address: 165 Allandale Road
City: Boston   **State:** MA   **Zip:** 02130

Telephone: (617) 524-7799   **Fax:** (617) 524-2825
E-mail: mherrera@msh.org
URL: http://www.msh.org

Other locations: Newton, MA; Arlington, VA; Copenhagen, Denmark; and throughout Africa, Latin America, and Asia
Founded: 1971

Mission:
MSH works with policymakers, managers, and health care providers to narrow the gap between knowledge and public health management action in order to improve the health and well-being of those disadvantaged by ameliorable risks.
MSH seeks to influence public policy; to improve management of health and family-planning services through increased availability, efficiency, effectiveness, and sustainability; and to promote access to these services.

Programs:
MSH activities focus on: the application of practical management skills to public health problems through both the public and private sectors, the extension of technical and management competence to individuals and institutions through collaborative work and training programs, innovation in applied research and health management, and the dissemination of experience through a number of publications and conferences.

Atmosphere: Informal, fast-paced, collegial, committed

Volunteers: 0

STAFF POSITIONS
Full-time staff: 160   **Part-time staff:** 25
Staff positions available: Depends on current contracts
Staff salary: $18,000 per year, comprehensive benefits to start

INTERNSHIPS
How Many: Varies   **Duration:** About 3 months
Remuneration: Unpaid
Duties:
Duties vary according to scope of work. These may include research, data analysis, writing, editing, project development, as well as general administrative support.

Qualifications:
Motivation, enthusiasm, integrity, ability to work independently and responsibility with a seriousness of purpose. Also health, pharmacy, finance/economics, or policy background is helpful.

How to apply: Cover letter and resume

Matsunaga Institute for Peace at the University of Hawaii

Address: University of Hawaii, 2424 Maile Way, Porteus 717
City: Honolulu   **State:** HI   **Zip:** 96822

Telephone: (808) 956-7427   **Fax:** (808) 956-5708
E-mail: uhip@hawaii.edu
URL: http://www2.hawaii.edu/uhip/

Other locations: None listed

Founded: 1986

Mission:
The Matsunaga Institute for Peace is an academic community designed to explore, develop, and share knowledge of peace through teaching, research, publication, and public service. The Institute promotes peace personally, locally, nationally, and globally through nonviolent and compassionate means. By addressing issues of conflict management, community building, and reduction of violence, the Institute approaches its goal: a world at peace.

Programs:
Peace Studies Major—individually designed via Liberal Studies.
Peace Studies Certificate—15-credit program in peace issues for undergraduate and classified graduate students.
Pacific Peace Seminar—summer interdisiciplinary courses.
Symposia and Lectures—diverse opinions on peace issues.
Community Outreach—programs of special relevance to Hawaii.
Resource Center—peace and conflict resolution publications.
Center for Global Nonviolence—solutions without force.
International Center for Democracy—democratic values, practices.

Atmosphere: Formal, casual, hectic at times

Volunteers: 0

STAFF POSITIONS
Full-time staff: 6   **Part-time staff:** 1
Staff positions available: None
Staff salary: $26,000 to $50,000 per year

INTERNSHIPS
How Many: Varies   **Duration:** Varies
Remuneration: Unpaid
Duties:
Computer and Internet work, special events planning, curriculum development.
Qualifications:
Independent, computer-literate, an interest in peace and conflict resolution issues.
How to apply: Cover letter and resume; phone calls welcome

Meadowcreek

Address: PO Box 100
City: Fox   **State:** AR   **Zip:** 72051

Telephone: (501) 363-4500   **Fax:** (501) 363-4578
E-mail: Not listed
URL: Not listed

Other locations: None listed

Founded: 1979

Mission:
Meadowcreek's goal is to promote lifestyles that exist in harmony with natural resources and environments. Meadowcreek supports healthy and meaningful lives for today's and future generations by exploring and demonstrating alternative energy technologies and natural gardening techniques. Publications include *Education for A Sustainable World: Environmental Activities for Junior High Students*, resource lists, and technical briefs on issues of interest.

Programs:
Meadowcreek is a nonprofit, environmental education and retreat center providing programs for school and business groups and the general public interested in learning ways to lead more sustainable lifestyles. Educational sessions include: garden tour, energy tour, environmental overview, cooperative games, and hikes. Meadowcreek's workshop series is especially valuable for persons striving for small-scale self-sufficiency. Sample topics for these programs include: beekeeping, solar electrical, low-maintenance landscaping, solar water heating, and healthy cooking.

Atmosphere: Casual, friendly, open, busy

Volunteers: Varies

STAFF POSITIONS
Full-time staff: 7   **Part-time staff:** 7 to 9
Staff positions available: 1 per year

Staff salary: Varies, benefits included

INTERNSHIPS
How Many: Varies   **Duration:** 3 to 9 months
Remuneration: $500 monthly, housing costs
Duties:
Gardening/landscaping, energy-related projects, education and tours, program coordination, publicity, office and educational assistance.

Qualifications:
Good communication skills, willingness to live and work in a rural area. Some background or general understanding of related topics preferred.

How to apply: Cover letter, resume, and application

JOBS YOU CAN LIVE WITH 1996

Medical Technology and Practice Patterns Institute (MTPPI)

Address: 2121 Wisconsin Avenue, NW, Suite 220
City: Washington **State:** DC **Zip:** 20002

Telephone: (202) 333-8841 **Fax:** (202) 333-5586
E-mail: inquiry@mt-ppi.org
URL: Not listed

Other locations: None listed

Founded: 1986

Mission:
MTPPI is dedicated to research, cooperation, education, and the exchange of information regarding new and emerging medical technologies.

Programs:
The Institute conducts research on the clinical, economic, and societal implications of health-care technologies and has recently been designated a World Health Organization Collaborating Center for Health Technology Assessment. MTPPI's research is directed toward the formulation and implementation of local, national, and international health-care policies. In the international arena, special attention is given to developing countries.

Atmosphere: Casual, smooth, relaxed

Volunteers: 0

STAFF POSITIONS
Full-time staff: 18 **Part-time staff:** 3
Staff positions available: Varies

Staff salary: Depends on experience, educational level, and skills

INTERNSHIPS
How Many: Not listed **Duration:** Not listed
Remuneration: None listed
Duties:
Not listed
Qualifications:
Not listed

How to apply: Cover letter and resume

Metasystems Design Group, Inc.

Address: 2000 N. 15th Street, Suite 103
City: Arlington   **State:** VA   **Zip:** 22201

Telephone: (703) 243-6622   **Fax:** (703) 841-9798
E-mail: abby@tmn.com
URL: http://www.tmn.com/

Other locations: None listed

Founded: 1983

Mission:
Metasystems started with a vision of using information technologies to enhance organizations, empower people, and contribute to "closing the gap between the human condition and the human potential." By promoting candor, cooperation, and creativity in on-line communication, Metasystems helps leaders understand the new technologies and bring the advantages of these tools to their organizations in ways that support their strategic objectives.

Programs:
One of Metasystem's major programs is the Meta Network which is a large virtual community of networks. Metasystems also concentrates on the design and publication of their Web pages, along with consulting in organizational development in similar issues.

Atmosphere: Casual, friendly, hectic

Volunteers: 0

STAFF POSITIONS
Full-time staff: 8   **Part-time staff:** 2
Staff positions available: 2 to 3 per year
Staff salary: Varies

INTERNSHIPS
How Many: Varies   **Duration:** 6 months to 2 years
Remuneration: Some stipends available
Duties:
Customer support, participating in on-line conferences, administrative tasks, and training users on the World Wide Web.
Qualifications:
Good communication and writing skills, a sense of humor.

How to apply: Cover letter and e-mail

Mid-South Peace and Justice Center

Address: PO Box 11428
City: Memphis   **State:** TN   **Zip:** 38111

Telephone: (901) 452-6997   **Fax:** (901) 452-7029
E-mail: mspjc@igc.org
URL: Not listed

Other locations: None listed

Founded: 1982

Mission:
The Mid-South Peace and Justice Center seeks to educate and empower individuals and groups, utilizing a variety of personal talents and collective means. The Center promotes participatory change in national, state, and local policies in a common spirit of nonviolence.

Programs:
The Center focuses on disarmament and federal spending priorities, local housing and banking issues, environmental issues, Central America, and Southern Africa. These issues are addressed by staff and volunteer task forces. The Center keeps a library, audio-visual resource center, research files, and a wide variety of free literature, which are all available to the public. The Center publishes a monthly newsletter, *Just Peace*, and produces a weekly television show also called *Just Peace* on Cablevision.

Atmosphere: Casual, friendly, hectic

Volunteers: Over 100

STAFF POSITIONS
Full-time staff: 2   **Part-time staff:** 2
Staff positions available: 1 per year
Staff salary: $15,000 to $17,000 per year, health insurance

INTERNSHIPS
How Many: No specific limit   **Duration:** Summer or semester
Remuneration: Unpaid

Duties:
Interns can focus on programs that best relate to their interests and are expected to help with general office tasks.

Qualifications:
Self-motivated, good communication and writing skills, and computer literacy is helpful.

How to apply: Phone call

JOBS YOU CAN LIVE WITH 1996

Midwest Renewable Energy Association

Address: PO Box 249
City: Amherst **State:** W I **Zip:** 54406

Telephone: (715) 824-5166 **Fax:** (715) 824-5399
E-mail: Not listed
URL: Not listed

Other locations: None listed

Founded: 1990

Mission:
The Midwest Renewable Energy Association promotes renewable energy and energy efficiency through education and demonstration.

Programs:
The Association's projects include:
Midwest Renewable Energy Fair—annual solstice celebration featuring workshops, exhibits, speakers, entertainment, and more.
ReNews—quarterly newsletter.
Renewable Energy and Energy Efficiency Directory—a directory of renewable energy and energy-efficiency professionals.

Atmosphere: Casual, relaxed July-April, hectic in May-June.

Volunteers: Hundreds

STAFF POSITIONS
Full-time staff: 1 **Part-time staff:** 2
Staff positions available: Varies
Staff salary: $12 per hour plus health insurance

INTERNSHIPS
How Many: 0 **Duration:** Not listed
Remuneration: Not listed
Duties:
Not listed

Qualifications:
Not listed

How to apply: Not listed

National Action Council for Minorities in Engineering (NACME)

Address: 3 W. 35 Street
City: New York **State:** NY **Zip:** 10003

Telephone: (212) 279-2626 ext. 23 **Fax:** (212) 629-5178
E-mail: Not listed
URL: Not listed

Other locations: None listed

Founded: 1974

Mission:
NACME's mission is to increase representation of African Americans, Latinos and American Indians in the nation's engineering profession. A nonprofit committed to excellence in education, NACME believes technical strength will determine world leadership into the new century, economic productivity rests on the quality of the scientists and engineers we produce, and our viability as a nation depends on our utilizing human resources from all segments of our population.

Programs:
Through highly leveraged partnerships with industry, government, and the academic community, NACME has developed strategies that effectively expand resources and yield significant return on investment. Research and public policy analysis are conducted in education; comprehensive student support is provided, including financial aid; model demonstration programs are developed, operated, and disseminated nationally through broadcasts, conferences, and publications.

Atmosphere: Corporate

Volunteers: Not listed

STAFF POSITIONS
Full-time staff: Not listed **Part-time staff:** Not listed
Staff positions available: Not listed
Staff salary: Not listed

INTERNSHIPS
How Many: Not listed **Duration:** Not listed
Remuneration: Not listed
Duties:
Not listed

Qualifications:
Not listed

How to apply: Not listed

National Association for Science, Technology, and Society (NASTS)

Address: 133 Willard Building
City: University Park   **State:** PA   **Zip:** 16802

Telephone: (814) 865-3044   **Fax:** (814) 865-3047
E-mail: ejbz@psu.edu
URL: Not listed

Other locations: None listed

Founded: 1988

Mission:
NASTS's goal is to provide a forum where people from every academic discipline—school teachers, scientists, engineers, professionals in religion, policymakers, and media— meet as equals to discuss, debate, and share concerns regarding society's handling of science and technology. NASTS seeks proactive guiding of technology and science according to the population's underlying values.

Programs:
Technological Literacy Conference—an annual national conference which involves members of the STS community in a weekend of lectures, workshops, and dialogue.
STS News—a quarterly newsletter that acts as a major communications link to members.
Bulletin of Science, Technology, and Society—a bimonthly magazine containing scholarly and reflective articles, book reviews, teaching aids, and a guide to periodical literature in STS issues.

Atmosphere: Hectic, stimulating, often fragmented, balanced by team effort

Volunteers: 0

STAFF POSITIONS
Full-time staff: Varies   **Part-time staff:** 2
Staff positions available: None

Staff salary: Varies

INTERNSHIPS
How Many: Not listed   **Duration:** Not listed
Remuneration: Not listed
Duties:
Not listed
Qualifications:
Not listed

How to apply: Not listed

National Audubon Society

Address: 666 Pennsylvania Avenue, SE, Suite 200
City: Washington   **State:** DC   **Zip:** 20003

Telephone: (202) 547-9009   **Fax:** (202) 547-9022
E-mail: Steve.Daigneault@Audubon.org
URL: http://www.Audubon.org/audubon

Other locations: Offices throughout the United States

Founded: 1905

Mission:
The National Audubon Society seeks to conserve and restore natural ecosystems, focusing on birds and other wildlife for the benefit of humanity and the earth's biological diversity.

Programs:
In the DC office, most of the work revolves around congressional activities. Recent projects focused on the Farm Bill, Clean Water Act, Endangered Species Act, refuge system issues, and population issues. Newsletters, action alerts, and fact sheets are also published.

Atmosphere: Casual

Volunteers: 4 to 7

STAFF POSITIONS
Full-time staff: 23   **Part-time staff:** 2
Staff positions available: 3 per year
Staff salary: $20,000 per year plus benefits for entry level

INTERNSHIPS
How Many: 4 to 7   **Duration:** 4 months
Remuneration: Little monetary, rewarding experience
Duties:
Write, edit, lobby, attend hearings, research, and perform administrative tasks.

Qualifications:
Passion, commitment, humility, personable, and strong writing and speaking skills.

How to apply: Cover letter, resume, and writing samples

National Center for Appropriate Technology (NCAT)

Address: PO Box 3838
City: Butte **State:** MT **Zip:** 59702

Telephone: (406) 494-4572 **Fax:** (406) 494-2905
E-mail: Not listed
URL: Not listed

Other locations: Fayetteville, AR; Austin, TX; Seattle, WA; and Sauk Rapids, MN
Founded: 1976

Mission:
The Center champions sustainable technologies and community-based approaches that protect natural resources and assist people, especially the economically disadvantaged, in becoming more self-reliant.

Programs:
Appropriate Technology Transfer for Rural Areas (ATTRA)—national sustainable agriculture information service designed for use by commercial farmers. ATTRA offers technical assistance and information to extension agents, agriculture support groups, researchers, educators, and agri-businesses, at no charge.
Low Income Home Energy Assistance Program (LIHEAP)—assist state and local program directors run their programs; provide information clearinghouse to program administrators; and help supplement federal money with nonfederal money to programs.

Atmosphere: Casual, relaxed, can be hectic, may be uncertainty in job longevity

Volunteers: 0

STAFF POSITIONS
Full-time staff: 29 **Part-time staff:** 39
Staff positions available: 1 to 2 per year
Staff salary: Depends on the position

INTERNSHIPS
How Many: 1 to 2 per year **Duration:** Varies
Remuneration: Hourly wage
Duties:
Depends on the position. It may include conducting surveys, gathering information over the telephone, setting up databases.
Qualifications:
Ability and willingness to learn new things; class coursework in the area of the internship.

How to apply: Cover letter and resume.

National Coalition Against the Misuse of Pesticides (NCAMP)

Address: 701 E Street, SE
City: Washington **State:** DC **Zip:** 20003

Telephone: (202) 543-5450 **Fax:** (202) 543-4791
E-mail: ncamp@igc.apc.org
URL: http://www.csn.net/ncamp/

Other locations: None listed

Founded: 1981

Mission:
NCAMP is a national membership organization of grassroots groups and individuals established to identify hazards of commonly used pesticides and to reduce or, where possible, eliminate unnecessary use of pesticides through the adoption of safe alternatives.

Programs:
NCAMP runs an information clearinghouse on pesticides and available alternatives, and is also producing a toolkit for people to evaluate pesticide use at the local level. NCAMP works on school issues and grassroots organizing through the Beyond Pesticides campaign. NCAMP publishes a quarterly news magazine, *Pesticides and You,* and a monthly news bulletin, *NCAMP's Technical Report.*

Atmosphere: Casual, hectic

Volunteers: 2

STAFF POSITIONS
Full-time staff: 3 **Part-time staff:** 4
Staff positions available: 0 to 1 per year
Staff salary: Varies widely

INTERNSHIPS
How Many: 4 per semester **Duration:** Semester
Remuneration: Unpaid
Duties:
NCAMP tries to meet the needs of both the organization and the intern. This means that an intern should be willing to help with the information clearinghouse, as well as work on a specific project.

Qualifications:
Interest in the issue of pesticides and some background in related fields.

How to apply: Cover letter, resume, and writing sample

National Council for International Health

Address: 1701 K Street, NW, Suite 600
City: Washington **State:** DC **Zip:** 20006

Telephone: (202) 833-5900 ext. 207 **Fax:** (202) 833-0075
E-mail: NCIH@cais.com
URL: Not listed

Other locations: None listed

Founded: 1971

Mission:
The mission of the National Council for International Health is to improve health worldwide by providing vigorous leadership and advocacy designed to increase the US public and private sector awareness of international health needs.

Programs:
NCIH work includes: communications, membership services, international partnerships and networking, AIDS programs, and education for international health professionals. Their publications include *Health Link*, *Career Network*, and *AIDS Link*.

Atmosphere: Casual

Volunteers: 3

STAFF POSITIONS
Full-time staff: 11 **Part-time staff:** 3
Staff positions available: 3 per year
Staff salary: $19,000 per year

INTERNSHIPS
How Many: 4 **Duration:** 3 to 6 months
Remuneration: Unpaid

Duties:
Research, editorial, administrative, and coordinating the annual job fair and conference.

Qualifications:
B.S. or M.S., attention to detail, computer literacy, and being a team player.

How to apply: Cover letter and resume; phone calls welcome

National Network of Minority Women in Science (MWIS)

Address: 1333 H Street, NW, Room 1130
City: Washington   **State:** DC   **Zip:** 20005

Telephone: (202) 326-6757   **Fax:**
E-mail: ggilbert@aaas.org
URL: None listed

Other locations: None listed

Founded: 1978

Mission:
As a part of AAAS, MWIS strives to provide minority female students with full access to career information and educational opportunities, and to promote the professional advancement of minority women scientists and engineers. The Network provides a communication and support system, and utilizes the scientific and technical knowledge of its members to enhance understanding of educational policymaking.

Programs:
MWIS has local and regional chapters that function as professional associations, produces a quarterly newsletter, and collects and disseminates information through other AAAS publications.

Atmosphere: None Listed

Volunteers: Not listed

STAFF POSITIONS
Full-time staff: 3   **Part-time staff:** None Listed
Staff positions available: Not listed
Staff salary: Not listed

INTERNSHIPS
How Many: Not listed   **Duration:** Not listed
Remuneration: Not listed
Duties:
Not listed

Qualifications:
Not listed

How to apply: Not listed

National Organization for Women (NOW)

Address: 1000 16th Street, NW, Suite 700
City: Washington **State:** DC **Zip:** 20036

Telephone: (202) 331-0066 **Fax:** (202) 785-8576
E-mail: now@now.org
URL: Not listed

Other locations: 600 chapters around the country

Founded: 1966

Mission:
NOW strives to eliminate discrimination and harassment in the workplace; secure abortion and birth-control rights for all women; stop all forms of violence against women; eradicate racism, sexism, and heterosexism; and promote equality and justice through mass actions, intensive lobbying, nonviolent civil disobedience, direct action, and litigation.

Programs:
NOW's programs deal with a variety of issues concerning women's rights and other subjects listed in the mission. They also publish *The National NOW Times* newspaper.

Atmosphere: A fast-paced, feminist environment

Volunteers: 20

STAFF POSITIONS
Full-time staff: 3 **Part-time staff:** 0
Staff positions available: 1 per year
Staff salary: Varies

INTERNSHIPS
How Many: 10 to 15 per semester **Duration:** 8 to 14 weeks
Remuneration: Unpaid
Duties:
Varied according to political climate. All are project-based, as opposed to clerical or administrative.

Qualifications:
Hard-working, feminist, flexible, good communication skills.

How to apply: Cover letter, resume, application, and reference letters

National Peace Foundation

Address: 1835 K Street, NW, Suite 610
City: Washington **State:** DC **Zip:** 20006

Telephone: (202) 223-1770 **Fax:** (202) 223-1718
E-mail: NPIFNATL@igc.apc.org
URL: Not listed

Other locations: None listed

Founded: 1982

Mission:
The Foundation's overall mission is to promulgate peace building and conflict resolution on every level, from the community to the regional, to the national and international.

Programs:
The National Peace Foundation distributes a periodic newsletter *Peace Reporter*, and is currently publishing a directory for international conflict resolution programs to be completed in 1996. The National Peace Foundation also works to integrate conflict-resolution education in US schools. There are also two major international training programs which are the Transcaucasus Women's Dialogue and a variety of Russian Civic and Political Forums.

Atmosphere: Relaxed

Volunteers: 0

STAFF POSITIONS
Full-time staff: 5 **Part-time staff:** 1
Staff positions available: Usually none
Staff salary: Varies

INTERNSHIPS
How Many: Usually 1 **Duration:** 3 to 6 months
Remuneration: Varies
Duties:
Some receptionist duties along with working on a specific project.
Qualifications:
Intelligent, quick learner, friendly, and interested in peace and conflict resolution.
How to apply: Cover letter, resume, and writing sample

National Recycling Coalition (NRC)

Address: 1727 King Street, Suite 105
City: Alexandria **State:** VA **Zip:** 22314-2720

Telephone: (703) 683-9025 **Fax:** (703) 683-9026
E-mail: Not listed
URL: Not listed

Other locations: None listed

Founded: 1978

Mission:
The NRC, a nonprofit organization committed to maximizing recycling as an integral part of solid waste and resource management, includes businesses, environmental and recycling organizations, state and local governments, and individuals. The NRC provides technical education, disseminates public information, shapes public and private policy on recycling, and operates programs fostering recycling to conserve resources and reduce waste.

Programs:
NRC's Recycling Advisory Council participated in the development and launch of the Chicago Board of Trade Recyclable Exchange; initiated 24 individuals into Recycling to Build Community, a joint program of NRC and AmeriCorps VISTA, which works in urban and rural low-income communities around the country in an effort to create economic development opportunities and jobs in recycling. Publications: bimonthly membership newsletter the *NRC Connection*, the Buy Recycled Business Alliance Newsline, and the *Market Development NewsLink*. Also available are buy recycled fact sheets and industry-specific guidebooks.

Atmosphere: Professional, casual, active

Volunteers: Varies

STAFF POSITIONS
Full-time staff: 17 **Part-time staff:** 1
Staff positions available: Varies
Staff salary: Depends on related experience

INTERNSHIPS
How Many: Varies **Duration:** 3 to 6 months
Remuneration: 1 paid position, $1000 monthly
Duties:
Researching recycling technologies, database management, and planning workshops.

Qualifications:
College degree, computer skills, interest in recycling technology.

How to apply: Cover letter, resume, and writing sample

National Security and Natural Resources News Service

Address: 1730 Rhode Island Avenue, NW, Suite 102
City: Washington **State:** DC **Zip:** 20036

Telephone: (202) 466-4510 **Fax:** (202) 466-4344
E-mail: Not listed
URL: Not listed

Other locations: None listed

Founded: 1990

Mission:
National Security and Natural Resources News Service works to increase and improve the major news media coverage of arms control, military and national security stories, natural resources, and the environment.

Programs:
Work focuses on exclusive and investigative stories that are then shown on the major television networks as well as in major newspapers and other publications.

Atmosphere: Professional office with a varying atmosphere

Volunteers: 4

STAFF POSITIONS
Full-time staff: 8 **Part-time staff:** 0
Staff positions available: None usually
Staff salary: Varies

INTERNSHIPS
How Many: 6 **Duration:** Semester
Remuneration: Unpaid
Duties:
Interns will work closely with the reporters and bureau chief on the research and development of the stories.

Qualifications:
Excellent research and communication skills; published work is a plus.

How to apply: Cover letter, resume, and writing sample

National Security Archive

Address: 2130 H Street, NW, Gelman Library, Suite 701
City: Washington **State:** DC **Zip:** 20037

Telephone: (202) 994-7000 **Fax:** (202) 994-7005
E-mail: nsarchiv@gwis2.circ.gwu.edu
URL: Not listed

Other locations: None listed

Founded: 1985

Mission:
The National Security Archive is an independent, nongovernmental research institute and library located at the George Washington University. The Archive collects and publishes declassified documents obtained through the Freedom of Information Act.

Programs:
The Archive has the world's largest nongovernmental collection of documents released through the FOIA and has established an international reputation as the most successful nonprofit user of the FOIA. The Archive has set many important precedents, including less burdensome qualifications for waivers of processing fees and the long-term preservation of computer tapes from the Reagan, Bush, and Clinton White House staffs. The Archive has gained the release of thousands of previously classified documents, such as the historic correspondence between Kennedy and Khrushchev during the Cuban Missile Crisis and Oliver North's notebooks.

Atmosphere: Casual, busy

Volunteers: 0

STAFF POSITIONS
Full-time staff: 18 **Part-time staff:** 10
Staff positions available: None

Staff salary: $19,000 to $30,000 per year

INTERNSHIPS
How Many: 10 to 12 per semester **Duration:** 2 months
Remuneration: Commuting expenses and academic credit
Duties:
Each intern works on a specific research project where assignments include: building chronologies of events, helping obtain and catalog government documents, assisting with data entry, and performing library research.

Qualifications:
Interns should have related interests, and should be in college, recently graduated, or in graduate school.

How to apply: Cover letter, resume, writing samples, academic transcript, and (optional) two recommendations

National Wildlife Federation

Address: 1400 16th Street, NW
City: Washington   **State:** DC   **Zip:** 20036-2266

Telephone: (202) 797-6800   **Fax:** (202) 797-6677
E-mail: Not listed
URL: http://www.nwf.org/nwf

Other locations: Vienna, VA; Anchorage, AK; Ann Arbor, MI; Montpelier, VT; Atlanta, GA; Portland, OR; Boulder, CO; Missoula, MT; Bismarck, ND; and Austin, TX
Founded: 1936

Mission:
NWF's mission is to educate, inspire, and assist individuals and organizations of diverse cultures to conserve wildlife and other natural resources and to protect the earth's environment in order to achieve a peaceful, equitable, and sustainable future.

Programs:
NWF's programs include: legislative lobbying, advocacy, activist organizing, environmental education programs for children and families, environmental litigation, corporate initiatives on the environment, campus sustainability, and the sale of nature education materials. Publications include: *National Wildlife*, *International Wildlife*, *Ranger Rick*, *Your Big Backyard*, *EnviroAction*, *Nature's Best*, and various activist guides, reports, and fact sheets.

Atmosphere: Casual and hectic

Volunteers: Over 100

STAFF POSITIONS
Full-time staff: 500   **Part-time staff:** 25 to 50
Staff positions available: Varies
Staff salary: Varies on experience

INTERNSHIPS
How Many: About 7 per office   **Duration:** 6 months
Remuneration: $275 per week plus benefits
Duties:
Research, attending congressional hearings and briefings, lobbying, grassroots organizing, conference planning, mailings, and general office support.

Qualifications:
Must be a college graduate.

How to apply: Cover letter, resume, writing sample, and references

National Women's Health Network

Address: 514 10th Street, NW, Suite 400
City: Washington   **State:** DC   **Zip:** 20004

Telephone: (202) 347-1140   **Fax:** (202) 347-1168
E-mail: Not listed
URL: Not listed

Other locations: None listed

Founded: 1976

Mission:
The Network provides women with information to lobbying federal agencies into making women's health concerns a top priority. As the only long-standing, public-interest organization dedicated exclusively to women and health, issues include: AIDS, reproductive rights, breast cancer, occupational and environmental health issues, older women's health, new contraceptive technologies, and national health-care reform.

Programs:
National Health Care Program—to work for a health care system more appropriate for all women.
Breast Health and Breast Cancer—to promote good breast care, lessen the number diagnosed with breast cancer, and advocate for the best detection methods and treatments.
Research on Women's Health—to remedy the under-representation of women in clinical research studies.
AIDS—to inform women about the necessity of prevention.
Reproductive Health—to enable all women to choose when, where, and with whom they bear children.

Atmosphere: Relaxed, busy

Volunteers: 5 to 8

STAFF POSITIONS
Full-time staff: 7   **Part-time staff:** 2
Staff positions available: 1 per year
Staff salary: Varies

INTERNSHIPS
How Many: 5 per semester   **Duration:** 3 months
Remuneration: Unpaid
Duties:
Interns will be involved in two activities: federal health policy projects and the National Women's Health Network Clearinghouse.

Qualifications:
An interest in related issues and a commitment to work 4 days a week for a minimum of 12 weeks.

How to apply: Cover letter, resume, and writing sample

Natural Resources Council of Maine

Address: 271 State Street
City: Augusta   **State:** ME   **Zip:** 04330

Telephone: (207) 622-3101   **Fax:** (207) 622-4343
E-mail: Not listed
URL: Not listed

Other locations: None listed

Founded: 1959

Mission:
The National Resources Council of Maine is dedicated to the protection, restoration, and preservation of Maine's environment.

Programs:
Programs focus on the issues of air quality, water toxics, forest protection, and other land issues. Publications include *Maine Environment* and *Traveling and Growing Species*.

Atmosphere: Casual dress, hectic

Volunteers: Several

STAFF POSITIONS
Full-time staff: 27   **Part-time staff:** 5
Staff positions available: Varies
Staff salary: Varies

INTERNSHIPS
How Many: 1 to 2   **Duration:** Summer or semester
Remuneration: Unpaid
Duties:
Research and writing.
Qualifications:
A commitment to the environment along with initiative and enthusiasm. Also an understanding of the administrative or legal process is helpful.

How to apply: Cover letter, resume, and optional writing sample

Natural Resources Defense Council (NRDC)

Address: 1350 New York Avenue, NW
City: Washington **State:** DC **Zip:** 20005

Telephone: (202) 783-7800 **Fax:** (202) 783-5917
E-mail: Not listed
URL: Not listed

Other locations: New York, NY; San Francisco, CA; and Los Angeles, CA

Founded: 1970

Mission:
The Natural Resources Defense Council is a national, nonprofit public-interest organization dedicated to protecting the global environment and preserving the earth's natural resources. NRDC is staffed by dozens of highly-trained and well-respected professionals including resource specialists, scientists, and more than 40 attorneys.

Programs:
Air and Energy—urban air pollution, ground level ozone, airborne toxics, acid rain, global warming, energy efficiency.
International Environment—multilateral bank assistance, energy, population, trade and the environment, UN reform.
Nuclear Weapons—environmental impacts of nuclear weapons production and dismantlement, high-level radioactive waste cleanup, disposal and treatment, arms reduction, testing.
Public Health—water contaminants and treatment, pollution.
Water—Clean Water Act, agricultural runoff, urban stormwater.
Land—ecosystem management, national forest and grazing reform.

Atmosphere: Varies

Volunteers: Varies

STAFF POSITIONS
Full-time staff: 40 **Part-time staff:** 1
Staff positions available: Varies
Staff salary: Varies

INTERNSHIPS
How Many: At least 4 **Duration:** About a semester
Remuneration: Legal interns, $400 per week
Duties:
Legal interns work with environmental court cases, and non-legal interns work on special projects along with administrative work.

Qualifications:
Legal interns: completed second year of law school, excellent research and writing abilities. Non-legal: experience in environment issues; language, grassroots organizing, and communication skills.

How to apply: Cover letter, resume, transcript, writing sample, and 3 references

Nature Conservancy

Address: 1815 N. Lynn Street
City: Arlington **State:** VA **Zip:** 22209

Telephone: (703) 841-5379 **Fax:** (703) 841-7292
E-mail: Not listed
URL: Not listed

Other locations: In all 50 states, Canada, the Pacific Rim, and Latin America

Founded: 1951

Mission:
The Nature Conservancy works to preserve plants, animals, and natural communities that represent the diversity of life on Earth by protecting the lands and waters they need to survive.

Programs:
Program areas include: science, stewardship, legal, international, government relations, marketing and communications, development, and administration. There is also *Nature Conservancy*, a bimonthly magazine distributed to members.

Atmosphere: Corporate, casual, fast-paced, deadline-driven, goal-oriented

Volunteers: Several thousand

STAFF POSITIONS
Full-time staff: 2000 **Part-time staff:** Varies
Staff positions available: Varies
Staff salary: Varies

INTERNSHIPS
How Many: 100 per year **Duration:** 2 to 3 months
Remuneration: Salary, sometimes housing
Duties:
Field surveying, manual labor, biological monitoring, and office work.

Qualifications:
Progress toward a related degree and an interest in conservation.

How to apply: Cover letter and resume

NGO Committee on Disarmament, Inc.

Address: 777 UN Plaza, #3B
City: New York   **State:** NY   **Zip:** 10017

Telephone: (212) 687-5340   **Fax:** (212) 687-1643
E-mail: disarmtimes@igc.apc.org
URL: http://www.peacenet.org/disarm/

Other locations: None listed

Founded: Early 1970's

Mission:
The Committee provides services to nongovernmental organizations (NGOs) involved in disarmament and peace work in the United Nations context, and provides information for diplomats, NGOs, and UN staff and opportunities for dialogue between them.

Programs:
NGO Committee on Disarmament hosts conferences at the UN and makes their transcripts available to the public and also conducts a program for networking among peace movement groups and NGOs worldwide. The Committee also publishes a newspaper titled *Disarmament Times*.

Atmosphere: Casual, busy, productive

Volunteers: Many

STAFF POSITIONS
Full-time staff: 1   **Part-time staff:** 1
Staff positions available: 0 to 1
Staff salary: Varies

INTERNSHIPS
How Many: 1 to 2 per semester   **Duration:** A semester to a year
Remuneration: Unpaid
Duties:
Besides some administrative work, it is largely up to the intern. This could include research, networking, or conference planning.

Qualifications:
An interest in international affairs, initiative, and computer skills are a plus.

How to apply: Cover letter and resume; phone calls welcome

Northern Alaska Environmental Center

Address: 218 Driveway Street
City: Fairbanks   **State:** AK   **Zip:** 99701

Telephone: (907) 452-5021   **Fax:** (907) 452-3100
E-mail: naec@polarnet.com
URL: http://www.igc.apc.org/refuge

Other locations: None listed

Founded: 1971

Mission:
The Center, with its 1,200 members, is dedicated to preserving wilderness and natural habitats in interior and northern Alaska and promoting conservation and sustainable use of its natural resources. The Center is neither a think tank nor a government agency, but a privately-funded, public-interest advocacy organization dedicated to empowering the grassroots through information, education, and promoting public participation in resource management decisions.

Programs:
Alaska Boreal Forest—organize activities, build coalitions with other groups and individuals concerning conservation, publish alerts for distribution, travel to villages to educate native people about impacts of logging on fish, and design and present a slide show on the sustainable uses and importance of the boreal forest.
Road Access—coordinate opposition to increased road access to Beaver Creek National Wild River and across wild Alaska.
Arctic National Wildlife Refuge—organize Refuge flights and grassroots events to protect the coastal plain from oil drilling, work with Fish and Wildlife Service to protect the wilderness and wildlife.

Atmosphere: Casual, busy at times

Volunteers: Over 20

STAFF POSITIONS
Full-time staff: 5   **Part-time staff:** 0
Staff positions available: Maybe 1

Staff salary: $7 to $8 per hour to start

INTERNSHIPS
How Many: Varies   **Duration:** At least two months
Remuneration: No stipend, housing possible
Duties:
Design/present slide shows, mobilize citizens on forest management, interact with agencies responsible for land management, produce action alerts, and educate activists of the opportunity and importance of conservation.

Qualifications:
Motivated, flexible, organized, responsible, writing skills, and familiarity with environmental thought.

How to apply: Cover letter, resume, 3 references, and writing sample

Nuclear Age Peace Foundation

Address: 1187 Coast Village Road, Suite 123
City: Santa Barbara **State:** CA **Zip:** 93108

Telephone: (805) 965-3443 **Fax:** (805) 568-0466
E-mail: napf@igc.apc.org
URL: Not listed

Other locations: None listed

Founded: 1982

Mission:
The Nuclear Age Peace Foundation is a nonprofit, nonpartisan, international organization that seeks to educate the public through its programs and publications, and provides leadership towards a nuclear weapons-free world under international law. Publications include: *Waging Peace Bulletin, Waging Peace Booklets*, and *Global Security Studies*.

Programs:
World Campaign to Abolish Nuclear Weapons—campaign to lobby for a nuclear-weapons-free world in the 21st century.
International Law Project—individual accountability under international law, establishment of an International Criminal Court.
Science for a Sustainable World Project—supports use of science for constructive rather than destructive purposes and encourages scientists to use their skills only for constructive ends.
Distinguished Peace Leadership Award—presented annually to an individual who has shown courageous leadership in the cause of peace.

Atmosphere: Congenial and busy

Volunteers: Varies

STAFF POSITIONS
Full-time staff: 2 **Part-time staff:** 3
Staff positions available: Varies
Staff salary: Varies

INTERNSHIPS
How Many: 3 to 5 **Duration:** Not listed
Remuneration: $6 per hour
Duties:
The internship program involves students in research projects that will have an ongoing educational benefit for other students and the broader population.
Qualifications:
Bright, committed, good research/writing skills, willingness to learn.

How to apply: Cover letter, resume, application, academic transcript, and writing sample

Overseas Development Network

Address: 333 Valencia Street, #330
City: San Francisco   **State:** CA   **Zip:** 94103

Telephone: (415) 431-4204   **Fax:** (415) 431-5953
E-mail: odn@igc.org
URL: Not listed

Other locations: None listed

Founded: 1983

Mission:
Overseas Development Network is a student-based organization that challenges and inspires students to address global development issues through education, action, and hands-on experience. ODN also produces a number of publications on the topics of international relations, international development, and educational issues.

Programs:
Development Education—sponsors informative activities with college chapters and helps campus communities face the challenges of international development.
Bike-Aid—an annual cross-country, 3,600-mile cycling adventure, combines global education, physical challenge, community service and outreach, fundraising, and personal growth, while allowing participants to work directly with grassroots organizations.
Partnership in Development—builds partnerships between students and community groups working toward sustainable, locally-initiated development.

Atmosphere: Casual

Volunteers: 3 to 5

STAFF POSITIONS
Full-time staff: 5   **Part-time staff:** 0.
Staff positions available: 0 to 1 per year
Staff salary: $19,000 per year

INTERNSHIPS
How Many: Varies   **Duration:** Varies
Remuneration: Unpaid
Duties:
Interns can assist with the production and outreach on ODN's numerous publications and assist with programs that match their specific area of interest.
Qualifications:
An interest in related issues and a dedication to working together for grassroots development throughout the world.

How to apply: Cover letter, resume, and application; phone calls welcome

Pacific Whale Foundation (PWF)

Address: Kealia Beach Plaza, 101 N. Kihei Road
City: Kihei   **State:** HI   **Zip:** 96753

Telephone: (808) 879-8860   **Fax:** (808) 879-2615
E-mail: pacwhale@igc.apc.org
URL: Not listed

Other locations: None listed

Founded: 1979

Mission:
The mission of the Pacific Whale Foundation is to educate the public, from a scientific perspective, about marine mammals and their ocean environment. PWF was begun by conservationist and marine biologist Gregory Kaufman. The Foundation has a membership with representation from every continent of the world and an internationally acclaimed reputation for increasing public efforts to protect the world's oceans.

Programs:
Educational efforts, including the Adopt-A-Whale program, whale watches and reef walks for schoolchildren, Ocean Van school presentations, volunteer field study, educational materials and curriculum development, and a marine mammal naturalist certification course are coordinated through the Ocean Outreach Program. PWF receives international recognition for educational marine excursions, and designs and produces public education materials for resource management agencies in the US and Australia. Efforts to highlight whales and dolphins resulted in recognition of Maui as a favored breeding area for humpback whales.

Atmosphere: Quite casual, but very hectic

Volunteers: Varies

STAFF POSITIONS
Full-time staff: 40   **Part-time staff:** 0
Staff positions available: 4 to 5 per year

Staff salary: Varies

INTERNSHIPS
How Many: 3   **Duration:** 10 to 12 days
Remuneration: Academic credit available
Duties:
Photographing whales and dolphins for identification, small boat handling, population modeling, acoustical recording and analysis, behavioral observation, and scientific report writing.

Qualifications:
Over 16 years old; motivation, commitment, patience, and physical energy to dedicate to the science of understanding and saving marine mammals.

How to apply: Application

Palouse-Clearwater Environmental Institute

Address: PO Box 8596, 112 W. 4th Street, Suite 1
City: Moscow **State:** ID **Zip:** 83843

Telephone: (208) 882-1444 **Fax:** (208) 882-8029
E-mail: pcei@moscow.com
URL: http://www.moscow.com/resources/pcei

Other locations: None listed

Founded: 1986

Mission:
The Palouse-Clearwater Environmental Institute's mission is to increase citizen involvement in decisions that affect our region's environment. Through community organizing and education, PCEI assists members of our communities in making environmentally sound and economically viable decisions that promote a sustainable future.

Programs:
Sustainable Agriculture—sustainable agriculture working groups and agricultural options network in Montana, eastern Washington, and Idaho.
Water Quality—Paradise Creek and wetland restoration.
Transportation—vanpools from Lewiston, ID to Moscow, ID and the upkeep of pedestrian and bicycle paths.
Public Relations and Fundraising—community outreach that includes youth education on environmental issues.
The Institute also publishes a quarterly newsletter titled *The Environmental Newsletter*.

Atmosphere: Casual, relaxed, sometimes hectic

Volunteers: 6

STAFF POSITIONS
Full-time staff: 5 **Part-time staff:** 3
Staff positions available: 1 per year
Staff salary: Varies

INTERNSHIPS
How Many: 1 to 2 **Duration:** 3 months
Remuneration: Depends on available funding
Duties:
Depends on specific program.

Qualifications:
Ability to work alone or in groups, a self-starter, computer experience, and knowledge of environmental issues.

How to apply: Cover letter and resume

Peace Action

Address: 1819 M Street, NW, #420
City: Washington   **State:** DC   **Zip:** 20006

Telephone: (202) 862-9740   **Fax:** (202) 862-9762
E-mail: papcog@igc.apc.org
URL: Not listed

Other locations: State affiliates in 27 states

Founded: 1957

Mission:
Peace Action works to promote global security by stopping nuclear weapons proliferation and testing, halting weapons trafficking, and promoting new federal spending priorities.

Programs:
Peace Action's major programs include the Peace Economy Campaign; the Arms Trade Campaign; the Nuclear Disarmament Campaign; and the Grassroots and Campus Network Building Program. Peace Action also produces a quarterly newsletter.

Atmosphere: Casual, hectic

Volunteers: 2 to 3

STAFF POSITIONS
Full-time staff: 13   **Part-time staff:** 2
Staff positions available: 1 to 2 per year
Staff salary: $20,000s per year plus health benefits

INTERNSHIPS
How Many: 2 to 3   **Duration:** 3 months minimum
Remuneration: $50 per week stipend
Duties:
Interns assist in research, writing, contacting Congress, producing and disseminating educational resources, mobilizing activists, attending coalition meetings, with administrative duties.

Qualifications:
Interns should have a strong interest in international peace and justice issues.

How to apply: Cover letter, resume, application, writing sample, and references

Peace and Justice Center

Address: 21 Church Street
City: Burlington   **State:** VT   **Zip:** 05401

Telephone: (802) 863-2345   **Fax:** (802) 863-2532
E-mail: pjc@together.org
URL: Not listed

Other locations: None listed

Founded: 1979

Mission:
The Peace and Justice Center works for a just, peaceful, and ecologically healthy world through education, advocacy, training, and nonviolent action.

Programs:
Peace and Justice News—a newsletter published 10 times per year.
VT Grassroots Directory—a listing of 290 social change groups in Vermont.
Racial Justice and Equity Project—advocacy for people of color in Vermont.
People's Campaign for Economic Democracy—mobilizing Vermonters against Contract on America.

Atmosphere: Casual, relaxed, at times hectic

Volunteers: 100

STAFF POSITIONS
Full-time staff: 5   **Part-time staff:** 2
Staff positions available: None
Staff salary: $9 per hour and health insurance

INTERNSHIPS
How Many: 1 to 2 per year   **Duration:** Varies
Remuneration: Unpaid
Duties:
Depends on the project and skills/interests of intern.
Qualifications:
Self-motivated, responsible, enthusiastic, and some background in racial and/or justice issues.

How to apply: Cover letter and resume

JOBS
YOU CAN
LIVE WITH
1996

Pesticide Action Network North America

Address: 116 New Montgomery, Suite 810
City: San Francisco   **State:** CA   **Zip:** 94105

Telephone: (415) 541-9140   **Fax:** (415) 541-9253
E-mail: panna@panna.org
URL: http://www.panna.org/panna/

Other locations: United Kingdom; Columbia; Senegal; and Malaysia

Founded: 1984

Mission:
The Pesticide Action Network North America Regional Center advocates adoption of ecologically sound practices in place of pesticide use. The Network believes that citizen action is essential to challenge global proliferation of pesticides, to defend basic rights to promote health and environmental quality, and to ensure the transition to a just and viable society.

Programs:
Pesticide Action Network North America works with over 90 affiliated consumer, labor, farm, public health, environment, and advocacy groups in Canada, Mexico, and the US, as well as with partners around the world, to demand that development agencies and governments redirect support from pesticides to safe alternatives. Together these groups have documented the disastrous effects of pesticide dependence on economies, justice, environment, and health; promoted and demonstrated sustainable agriculture; and achieved changes in pest management policies and practices.

Atmosphere: Busy, informal

Volunteers: 2

STAFF POSITIONS
Full-time staff: 10   **Part-time staff:** 0
Staff positions available: 1 per year
Staff salary: $20,000 to $31,000 per year

INTERNSHIPS
How Many: 1 to 2   **Duration:** Varies
Remuneration: Local transportation costs
Duties:
Usually a combination of research, administrative, writing, and on-line communications.

Qualifications:
A degree or coursework in a related field—agriculture, international policy studies, public health, etc.

How to apply: Cover letter, resume, and writing sample

Physicians for Human Rights (PHR)

Address: 100 Boylston Street, #702
City: Boston   **State:** MA   **Zip:** 02116

Telephone: (617) 695-0041   **Fax:** (617) 695-0307
E-mail: phrusa@igc.apc.org
URL: gopher://gopher.humanrights.org:5000

Other locations: San Francisco, CA and Chicago, IL

Founded: 1986

Mission:
PHR is an organization of health professionals, scientists, and concerned citizens that use the knowledge and skills of the medical and forensic sciences to investigate and prevent violations of international human rights and humanitarian law. PHR adheres to a policy of strict impartiality and is concerned with the health consequences of human rights abuses, regardless of the ideology of the offending government or group.

Programs:
Programs include: investigative missions; International Campaign to Ban Landmines; educational speakers and programs; a development program; and a medical, legal, and forensic human rights program. Publications include: newsletters and annual reports, medical action alerts, and country and issue specific reports.

Atmosphere: Committed, hard-working, casual, sometimes hectic

Volunteers: 4

STAFF POSITIONS
Full-time staff: 13   **Part-time staff:** 2
Staff positions available: 1 per year
Staff salary: Depends on position

INTERNSHIPS
How Many: 8 per year   **Duration:** 3 to 4 months
Remuneration: Unpaid

Duties:
Interns assist staff in the research of health and human rights issues, draft letters to foreign governments, research human rights violations in specific countries, assist in the organization of fact-finding missions, and conduct research on related issues as needed.

Qualifications:
Excellent writing skills, ability to work alone and on multiple projects, computer skills. Prefer experience in political science, international relations, anthropology, languages, or regional country studies.

How to apply: Cover letter, resume, and writing sample

Physicians for Social Responsibility (PSR)

Address: 1101 14th Street, NW, Suite 700
City: Washington   **State:** DC   **Zip:** 20005

Telephone: (202) 898-0150   **Fax:** (202) 898-0172
E-mail: psrnatl@igc.apc.org
URL: http://www.psr.org:8000/

Other locations: None listed

Founded: 1961

Mission:
PSR is a national, nonprofit membership organization of over 20,000 health professionals and supporters working to promote nuclear arms reduction, international cooperation, protection of the environment and the reduction of violence. PSR was founded in 1961 and is the US affiliate of International Physicians for the Prevention of Nuclear War, which was awarded the Nobel Peace Prize in 1985.

Programs:
PSR's activities include public education about nuclear weapons production facilities, the social costs of the arms race, and the links between pollution and public health. PSR's programs range from citizen advocacy with Congress, speaker tours, media work, and educational publications.

Atmosphere: Casual, relaxed

Volunteers: 2

STAFF POSITIONS
Full-time staff: 25   **Part-time staff:** 2
Staff positions available: 1 per year
Staff salary: $20,000 to $27,500 per year

INTERNSHIPS
How Many: 2   **Duration:** 3 to 6 months
Remuneration: $200 dollars per week
Duties:
Research, writing, daily office responsibilities, monitoring congressional hearings and briefings, and assisting with grassroots membership.

Qualifications:
A background in nuclear weapons and security issues, excellent communications skills, computer literacy, and knowledge of basic office skills.

How to apply: Cover letter, resume, and writing sample

Plugged In

Address: 1923 University Avenue
City: Palo Alto   **State:** CA   **Zip:** 94303

Telephone: (415) 322-1134   **Fax:** (415) 322-6147
E-mail: hectorc@pluggedin.org
URL: http://www.pluggedin.org

Other locations: None

Founded: 1992

Mission:
Plugged In is a community computer access organization that provides training and support locally as well as on a national basis through the Internet.

Programs:
Plugged In involves teens using the Internet to provide on-line services and start-up information to industry and nonprofit organizations. Plugged In is also part of TIIAP, which develops communications infrastructures in Palo Alto for the nonprofit and business sectors. Plugged In also produces a bimonthly newsletter that discusses related issues.

Atmosphere: Casual, very busy

Volunteers: 12

STAFF POSITIONS
Full-time staff: 9   **Part-time staff:** 2
Staff positions available: 1 to 2 per year
Staff salary: Varies

INTERNSHIPS
How Many: 4 to 5 per year   **Duration:** 4 to 8 weeks
Remuneration: Housing, academic credit
Duties:
Clerical, Web work, and teaching classes about the Internet.

Qualifications:
An interest in technical and educational issues, sensitivity to diversity issues, and an interest in working with low-income populations. Technical skills a plus but not required.

How to apply: Cover letter, resume, and application

Population Association of America

Address: 721 Ellsworth Drive, Suite 303
City: Silver Spring **State:** MD **Zip:** 20904

Telephone: (301) 565-6710 **Fax:** (301) 565-7850
E-mail: 74761.1510@compuserve.com
URL: Not listed

Other locations: Public Affairs Office in Washington, DC

Founded: 1931

Mission:
The Population Association of America is an educational, nonprofit professional membership organization.

Programs:
The Population Association puts out a flagship journal on demographic issues and a newsletter for members that discusses population issues. There is also a directory of members and an annual meeting for members to discuss relevant issues.

Atmosphere: Casual

Volunteers: 0

STAFF POSITIONS
Full-time staff: 2 **Part-time staff:** 0
Staff positions available: Not listed
Staff salary: Not listed

INTERNSHIPS
How Many: 0 **Duration:** Not listed
Remuneration: Not listed
Duties:
Not listed
Qualifications:
Not listed

How to apply: Not listed

Powder River Basin Resource Council

Address: Box 1178
City: Douglas   **State:** WY   **Zip:** 82633

Telephone: (307) 358-5002   **Fax:** (307) 358-5002
E-mail: Not listed
URL: Not listed

Other locations: Sheridan, WY

Founded: 1973

Mission:
Powder River Basin Resource Council is a grassroots organization of individuals and affiliate groups dedicated to good stewardship of Wyoming's natural resources. Formed in response to the rapid energy development of that period, the founders were ranchers and others who came together over their concerns about strip-mining. The goal is to foster responsible development, consistent with preservation of Wyoming's unique natural heritage and lifestyle.

Programs:
Programs revolve around: the preservation and enrichment of our agricultural heritage and rural lifestyle, the conservation of Wyoming's unique land, mineral water and clean air consistent with responsible use of these resources to sustain the livelihood of present and future generations, and the education and empowerment of our citizens to raise a coherent voice in the decisions that will impact Wyoming residents' environment and lifestyle.

Atmosphere: Casual, hectic, and professional

Volunteers: 10

STAFF POSITIONS
Full-time staff: 3   **Part-time staff:** 1
Staff positions available: Usually none
Staff salary: Starts at $15,000 per year plus full benefits

INTERNSHIPS
How Many: 1 per year   **Duration:** 3 months
Remuneration: Room and board
Duties:
Help with issue work and special projects.

Qualifications:
Good communication skills.

How to apply: Cover letter, resume, and writing sample

Public Citizen Critical Mass Energy Project

Address: 215 Pennsylvania Avenue, SE
City: Washington **State:** DC **Zip:** 20003

Telephone: (202) 546-4996 **Fax:** (202) 547-7392
E-mail: Not listed
URL: Not listed

Other locations: None listed

Founded: 1974

Mission:
Public Citizen's Critical Mass Energy Project was founded by Ralph Nader to oppose nuclear power and promote cleaner, safer energy alternatives.

Programs:
Critical Mass prepares and disseminates reports, lobbies Congress, and acts as a watchdog of key federal and state energy regulatory agencies. With other citizens' groups and individuals across the country, Critical Mass empowers people to participate in decisions affecting health, safety, and standard of living; supports decreasing US dependence on coal and oil; and looks to a sustainable energy future, with cleaner, safer, renewable technologies like solar and wind power and efficient energy use. Critical Mass is a critic of the atomic power industry, which promised energy "too cheap to meter" and delivered high cost, accidents, and mounting radioactive waste.

Atmosphere: Casual, busy, hectic

Volunteers: 0

STAFF POSITIONS
Full-time staff: 5 **Part-time staff:** 1
Staff positions available: 1 per year
Staff salary: Varies

INTERNSHIPS
How Many: 1 to 3 **Duration:** Semester
Remuneration: Possibly commuting expenses
Duties:
Research and writing on energy-related policy and legal matters. Lobbying, working with the media, and performing administrative tasks.

Qualifications:
Experience and interest in science and alternative energy programs. Hard-working and intelligence are important qualities.

How to apply: Cover letter, resume, and writing sample

Public Voice for Food and Health Policy

Address: 1101 14th Street, NW, Suite 710
City: Washington   **State:** DC   **Zip:** 20005

Telephone: (202) 371-1840   **Fax:** (202) 371-1910
E-mail: pvoice@ix.netcom.com
URL: http://www.Publicvoice.org/pv

Other locations: None listed

Founded: 1982

Mission:
Public Voice for Food and Health Policy is an advocate for a safer, healthier, and more affordable food supply.

Programs:
Program areas include nutrition, food safety, sustainable agriculture, environmental factors, and inner city food access. Publications include the *Advocacy Update* newsletter and a variety of research reports.

Atmosphere: Informal, hectic when Congress is in session

Volunteers: 0

STAFF POSITIONS
Full-time staff: 15   **Part-time staff:** 1
Staff positions available: 1 to 2 per year
Staff salary: $25,000 to $50,000 per year, depending on position and experience

INTERNSHIPS
How Many: 2 to 3 per semester   **Duration:** 8 to 12 weeks, generally
Remuneration: Up to $200 per week for graduate students
Duties:
Research, writing, attending coalition meetings, Capitol Hill visits, and advocacy.

Qualifications:
Self-starter, good oral and written communication skills, computer literate, research-oriented, commitment to public interest work, and a willingness to explore new issues.

How to apply: Cover letter, resume, and writing sample

JOBS YOU CAN LIVE WITH 1996

Rainforest Action Network

Address: 450 Sansome Street, #700
City: San Francisco **State:** CA **Zip:** 94111

Telephone: (415) 398-4404 **Fax:** (415) 398-2732
E-mail: rainforest@ran.org
URL: http://www.ran.org/ran/

Other locations: Santa Monica, CA

Founded: 1985

Mission:
Rainforest Action Network works to protect the earth's rainforests and support the rights of their inhabitants through education, grassroots organizing, and nonviolent direct action.

Programs:
Major education and advocacy programs include: Amazon Program, Boycott Mitsubishi Campaign, Wood-Use Reduction Campaign, Rainforest Action Groups, International Information Clearinghouse, and Protect-An-Acre Program. Publications include *World Rainforest Week*, *World Rainforest Reports*, and *RAN Action Alerts*.

Atmosphere: Casual, hectic

Volunteers: 100 to 150

STAFF POSITIONS
Full-time staff: 22 **Part-time staff:** 10
Staff positions available: 4 to 5 per year
Staff salary: $20,000 per year, health/other benefits

INTERNSHIPS
How Many: 10 to 12 **Duration:** 3 months
Remuneration: Commuting expenses
Duties:
It depends on the specific project, but all interns have to devote 50% of their time toward clerical duties.

Qualifications:
Basic Macintosh computer skills, environmental knowledge, passion to save the environment, and a commitment of at least 12 hours per week for 3 months.

How to apply: Resume, application, and writing sample

Refugees International

Address: 21 Dupont Circle, NW
City: Washington **State:** DC **Zip:** 20036

Telephone: (202) 828-0110 **Fax:** (202) 828-0819
E-mail: ri@clark.net
URL: Not listed

Other locations: None listed

Founded: 1979

Mission:
Founded in response to forced repatriation of Cambodian and Vietnamese refugees, Refugees International provides early warning in crises of mass exodus. RI serves as the advocate of the unrepresented—refugees, displaced persons, and the dispossessed around the world. RI's mission is to bring the plight of refugees to global attention and to cajole, persuade, or embarrass governments and organizations that have the means—but not the will—to help.

Programs:
In recent years, Refugees International has moved from its initial focus on Indochinese refugees to global coverage, conducting almost 30 emergency missions in the last four years. They have answered the emergency calls of: Kurds stranded along the mountainous Turkish border; Burmese forced to flee to Bangladesh; war victims in Bosnia; Africans fleeing strife and famine in Liberia, Ethiopia, Somalia; and Rwandans surging to Tanzania and Zaire.

Atmosphere: Casual

Volunteers: 1

STAFF POSITIONS
Full-time staff: 10 **Part-time staff:** 1
Staff positions available: Varies
Staff salary: Not listed

INTERNSHIPS
How Many: 1 **Duration:** Varies
Remuneration: Commuting expenses
Duties:
Information collecting and meetings with other agencies.

Qualifications:
International background and resourcefulness.

How to apply: Cover letter, resume, and writing sample

Renew America

Address: 1400 16th Street, NW, Suite 710
City: Washington **State:** DC **Zip:** 20036

Telephone: (202) 232-2252 **Fax:** (202) 232-2617
E-mail: renewamerica@igc.apc.org
URL: http://solstice.crest.brg/environment/renew-

Other locations: None listed

Founded: 1979

Mission:
As America's leading source for environmental solutions, Renew America coordinates a network of community groups, environmental organizations, businesses, governmental leaders, and civic activists involved in environmental improvement. Renew America seeks out and promotes exemplary programs which offer positive, constructive models to inspire communities and businesses to meet environmental challenges.

Programs:
Environmental Success Index—chronicles projects across the country that measurably protect, restore, or enhance the environment.
National Awards for Environmental Sustainability—highlights about 25 programs every year that reached success in environmental improvement and that are working in communities to reduce solid waste and protect endangered species.
Annual Town Meeting—Renew America hosts an interactive national teleconference, linking national opinion makers with local communities facing common environmental challenges.

Atmosphere: Casual, busy

Volunteers: Varies

STAFF POSITIONS
Full-time staff: 6 **Part-time staff:** 0
Staff positions available: Varies
Staff salary: Varies

INTERNSHIPS
How Many: Varies **Duration:** 6 months
Remuneration: Stipend
Duties:
Much of the intern's time is spent working with local grassroots environmental organizations nationwide to identify projects for inclusion in the index. Interns will also assist with general activities of the programs and routine office work including computers.

Qualifications:
Must be a student with an interest in environmental issues.

How to apply: Cover letter, resume, and 2 to 3 references

Rhode Island Solar Energy Association

Address: 42 Tremont Street
City: Cranston   **State:** RI   **Zip:** 02920

Telephone: (401) 942-6691   **Fax:** Not listed
E-mail: Not listed
URL: Not listed

Other locations: None listed

Founded: 1976

Mission:
The Association educates the public and members about renewable energy through newsletters, workshops, seminars, tours, and references for the application and maintenance of renewal energy resources.

Programs:
The Association's programs include solar tours of homes and industries, various related workshops and seminars, and the publication of a quarterly newsletter titled *Helio*.

Atmosphere: Quite informal

Volunteers: 3

STAFF POSITIONS
Full-time staff: 3   **Part-time staff:** 0
Staff positions available: All volunteer

Staff salary: None

INTERNSHIPS
How Many: 0   **Duration:** Varies
Remuneration: Unpaid
Duties:
Not listed

Qualifications:
Not listed

How to apply: Phone call

River Watch Network (RWN)

Address: 153 State Street
City: Montpelier **State:** VT **Zip:** 05602

Telephone: (802) 223-3840 **Fax:** (802) 223-6227
E-mail: Not listed
URL: Not listed

Other locations: 13 states and in Canada, Mexico, and Hungary

Founded: 1987

Mission:
RWN brings people with disparate interests together to monitor, restore, and protect rivers. The goals are to define the issues concerning the river, and design and execute scientifically credible studies which: assess the condition of, and human impact on, river ecosystems; assess the river's ability to support both human uses and aquatic life; and use the studies' results to create appropriate strategies for river conservation through community action.

Programs:
RWN's programs work with the following rivers: the 12 major rivers in Vermont, Connecticut River, Hudson River Watershed, Merrimack River, Tribal Assistance Project, Mississippi River, Rio Grande, and the Danube River in Hungary. Publications include *Program Organizing Guide*, *Benthic Macroinvertebrate Monitoring Manual*, *River Sampling and Analysis Manual*, *Study Design Workbook*, and *High School Water Quality Study Guide*.

Atmosphere: Casual, relaxed, busy

Volunteers: 0

STAFF POSITIONS
Full-time staff: 4 **Part-time staff:** 2
Staff positions available: None
Staff salary: $25,000 to $30,000 per year

INTERNSHIPS
How Many: 1 to 3 **Duration:** Varies
Remuneration: Usually unpaid
Duties:
Community organizing, technical, administrative, and many others.

Qualifications:
Enthusiasm, commitment, and good communication skills.

How to apply: Cover letter and resume; phone calls welcome

Riveredge Nature Center

Address: PO Box 26
City: Newburg   **State:** W I   **Zip:** 53060

Telephone: (414) 375-2715   **Fax:** (414) 375-2714
E-mail: tc@omnifest.uwm.edu
URL: Not listed

Other locations: None listed

Founded: 1969

Mission:
The mission of Riveredge is to provide leadership in environmental education, to preserve its natural sanctuary, and to serve as a regional resource for scientific research. By educating people of all ages about their interrelationship with the earth and its plant and animal life, the Center fosters responsible environmental decisionmaking.

Programs:
Programs include: Testing the Waters, which involves 12,000 middle and high school students annually; joint venture with International Crane Foundation; exchange program with Russian and Riveredge naturalists; ongoing studies of species counts and endangered species populations; Prairie Invertebrate Conference to study the ecology of the prairie ecosystem; adventuresome summer camps for all ages; custom adult excursions within and beyond Wisconsin's borders; hands-on classes and imaginative family events, including Halloween Haunted Hike, ECOFOCUS, and Harvest Fest.

Atmosphere: Casual, professional, busy, family-like

Volunteers: 500

STAFF POSITIONS
Full-time staff: 7   **Part-time staff:** 7
Staff positions available: 1 per year, many volunteer positions

Staff salary: $18,000 to $27,000 per year depending on education and experience

INTERNSHIPS
How Many: 0   **Duration:**
Remuneration:
Duties:
Not listed

Qualifications:
Not listed

How to apply: Not listed

JOBS YOU CAN LIVE WITH 1996

Rocky Mountain Institute

Address: 1739 Snowmass Creek Road
City: Snowmass   **State:** CO   **Zip:** 81654-9199

Telephone: (970) 927-3851   **Fax:** (970) 927-3420
E-mail: orders@rmi.org
URL: None listed

Other locations: None listed

Founded: 1982

Mission:
Rocky Mountain Institute is an independent, nonprofit research and educational foundation with a vision across boundaries. The Institute's mission is to foster the efficient and sustainable use of resources as a path to global security.

Programs:
The Rocky Mountain Institute focuses its research on several interlinked areas including energy, transportation, real-estate development, water and agriculture, economic sustainability, and security. All programs are supported by vigorous outreach using the media.

Atmosphere: Casual, relaxed, located in a rural setting

Volunteers: 0

STAFF POSITIONS
Full-time staff: 42   **Part-time staff:** 2
Staff positions available: 1 to 4 per year
Staff salary: Varies

INTERNSHIPS
How Many: Depends on funding   **Duration:** Varies
Remuneration: No set salary
Duties:
Each of the seven programs would offer a great deal of variety for college students specializing in energy, transportation, green development services, water, agriculture, global security, and economic renewal.

Qualifications:
At least two years of college in a specialized field relating to one of the programs.

How to apply: Cover letter, resume, informal academic transcript, and writing sample

Search for Common Ground

Address: 1601 Connecticut Avenue, NW
City: Washington **State:** DC **Zip:** 20009

Telephone: (202) 265-4300 **Fax:** (202) 232-6718
E-mail: Not listed
URL: Not listed

Other locations: Ukraine, St. Petersburg, Moscow, Burundi, Macedonia, and Gaza
Founded: 1982

Mission:
Search for Common Ground is an independent, nonprofit organization dedicated to transforming conflict into cooperative action. Programs are designed to find workable solutions to divisive national and international problems, and the goal is to build a more secure and peaceful world. Search for Common Ground strives to be on the cutting edge as social innovators with contentious public policy issues.

Programs:
Major programs include the Network For Life and Choice (USA) and other similar issues in Macedonia, Burundi, Angola, Russia, Ukraine, and the Middle East.

Atmosphere: Relaxed, busy, informal, casual, intense

Volunteers: 0

STAFF POSITIONS
Full-time staff: 25 **Part-time staff:** 3
Staff positions available: Varies
Staff salary: Varies

INTERNSHIPS
How Many: 6 **Duration:** Varies, up to 1 year
Remuneration: Varies
Duties:
Research, organizing, writing, administrative, and project involvement.

Qualifications:
Bright, regional expertise, functional expertise, team player, hard worker, and sense of humor.

How to apply: Cover letter and resume

Shaver's Creek Environmental Center

Address: The Pennsylvania State University, 508A Keller Bldg.
City: University Park   **State:** PA   **Zip:** 16802-9976

Telephone: (814) 863-2000   **Fax:** (814) 865-2706
E-mail: djw105@psu.edu
URL: Not listed

Other locations: None listed

Founded: 1976

Mission:
Shaver's Creek Environmental Center extends the University's instruction, service, and research mission into the community by teaching and modeling of knowledge, values, skills, and experiences. The Center's dedication enables individuals and communities to achieve and maintain harmony between human activities and the natural systems that support all living species.

Programs:
Administered by Penn State's Division of Continuing and Distance Education, the center provides facilities and programming to meet education and research needs. The Center offers year-round environmental education programs for group visits; natural/cultural history exhibits; live amphibians and reptiles; hiking trails, herb gardens. The Raptor Center is federally and state licensed, providing perpetual care and housing for eagles, falcons, hawks, and owls. Shaver's Creek is a member of the Global Network of Environmental Education Centers and is involved in fostering environmental programming in central Europe.

Atmosphere: Varies from relaxed to hectic

Volunteers: 40

STAFF POSITIONS
Full-time staff: 11   **Part-time staff:** 7
Staff positions available: 1 per year
Staff salary: $21,000 per year with full benefits

INTERNSHIPS
How Many: 6 per season   **Duration:** Semester, summer
Remuneration: Housing, $100 per week, Internet access, resume writing
Duties:
Interns lead natural and cultural history programs for school and community groups, families, and the general public. Interns contribute articles to the newsletter, lead adventure programs, participate in the care of the animals, and assist in the general operation of the center.

Qualifications:
Strong desire to teach, sharing their knowledge and enthusiasm for the natural world. Background in education or natural sciences preferred. Undergraduates and graduates encouraged to apply.

How to apply: Application; phone calls welcome

Sigma Xi, the Scientific Research Society

Address: PO Box 13975
City: Research Triangle Park **State:** NC **Zip:** 27709

Telephone: (919) 549-4691 **Fax:** (919) 549-0090
E-mail: efer@sigmaxi.org
URL: http://www.sigmaxi.org

Other locations: None listed

Founded: 1886

Mission:
The mission of Sigma Xi is to honor scientific accomplishments, to encourage and enhance the worldwide appreciation and support of original investigations in science and technology, and to foster worldwide creative and dynamic interaction among science, technology, and society.

Programs:
Sigma Xi sponsors an annual forum on issues that intersect science and society and publishes its proceedings. The Society sponsors various activities on science and technology policy, education, and science and society issues, including lectureships for its chapters and awards of over $400,000 per year to support student research. Many activities are coordinated through 511 chapters at academic, industrial, and government facilities. The Society is expanding its local groups outside North America, becoming a global society of scientists and engineers. The Society's award-winning magazine, *American Scientist*, is published six times per year.

Atmosphere: Casual and sometimes hectic

Volunteers: 0

STAFF POSITIONS
Full-time staff: 38 **Part-time staff:** 1
Staff positions available: 2 to 3 per year
Staff salary: Salary varies, all positions include full benefits

INTERNSHIPS
How Many: 0 **Duration:** Not listed
Remuneration: Not listed
Duties:
Not listed

Qualifications:
Not listed

How to apply: Not listed

Smithsonian Environmental Research Center (SERC)

Address: PO Box 28
City: Edgewater   **State:** MD   **Zip:** 21037

Telephone: (410) 798-4424   **Fax:** (301) 261-7954
E-mail: HADDON@SERC.SI.EDU
URL: http://WWW.SERC.SI.EDU

Other locations: None listed

Founded: 1965

Mission:
The Mission of the Smithsonian Environmental Research Center (SERC) is to increase our knowledge about the interactions of flora and fauna with their environment and to disseminate this knowledge to improve our stewardship of the biosphere.

Programs:
Research and education at SERC cuts across traditional academic disciplines to investigate ecological processes at a wide range of scales in time and space. SERC's interdisciplinary studies combine experimental and long-term synthetic approaches to measure, predict, and test knowledge of environmental linkages. Areas of research include: quantifying effects of atmospheric deposition and land use on ground water, streams, and estuaries; addressing complex landscapes and biological consequences of natural and human disturbances; and controlling and integrating populations, communities, and ecosystems at boundaries between land and sea.

Atmosphere: Casual

Volunteers: 40

STAFF POSITIONS
Full-time staff: 100   **Part-time staff:** 10
Staff positions available: 0 to 1 per year
Staff salary: GS-3-7 (federal government grades)

INTERNSHIPS
How Many: 12 to 15   **Duration:** 10 to 12 weeks
Remuneration: $190 per week
Duties:
A variety of field research experiences.

Qualifications:
Good GPA, dependable, shows initiative, and the ability to write and communicate.

How to apply: Application by e-mail, academic transcript, and writing sample

Social Action and Leadership School for Activists

Address: 1601 Connecticut Avenue, NW, Suite 500
City: Washington   **State:** DC   **Zip:** 20009

Telephone: (202) 234-9382 x 229   **Fax:** (202) 387-7915
E-mail: Not listed
URL: Not listed

Other locations: None listed

Founded: 1993

Mission:
The mission of the School is to provide skills-building classes and in-depth policy seminars to progressive organizations and individuals.

Programs:
Skills training is offered in the areas of fundraising, communications, organizational development, on-line activism, research, and organizing. Policy seminars examine domestic and global affairs in great detail, combining a theoretical foundation with policy critiques and policy applications.

Atmosphere: Somewhat casual, busy

Volunteers: Varies

STAFF POSITIONS
Full-time staff: 1   **Part-time staff:** Varies
Staff positions available: None
Staff salary: None

INTERNSHIPS
How Many: 1   **Duration:** 3 to 6 months
Remuneration: Currently unpaid
Duties:
Assist program director in marketing and public relations, and assist in program management.

Qualifications:
Communication skills, writing and editing skills, and computer literate.

How to apply: Cover letter, resume, and writing sample

Society of Environmental Journalists

Address: PO Box 27506
City: Philadelphia   **State:** PA   **Zip:** 19118

Telephone: (215) 247-9710   **Fax:** (215) 247-9712
E-mail: SEJoffice@AOL.COM
URL: http://www.tribnet.com/

Other locations: None listed

Founded: 1990

Mission:
The Society of Environmental Journalists seeks to advance public understanding of environmental issues by improving the quality, accuracy, and visibility of environmental reporting. The Society is a membership organization of 1,100 that includes journalists and students.

Programs:
Programs include: national and regional conferences, membership directory, a guide to on-line services, and a quarterly journal for members.

Atmosphere: Casual

Volunteers: Dozens

STAFF POSITIONS
Full-time staff: 2   **Part-time staff:** 2
Staff positions available: Usually, only part-time available
Staff salary: $7 per hour as a starting salary

INTERNSHIPS
How Many: 1 to 2   **Duration:** Summer or semester
Remuneration: A small stipend
Duties:
Research, correspondence, and nonprofit organization systems.
Qualifications:
A strong interest in journalism and the environment, a quick learner, easy to get along with, curious, good writing skills, and good telephone skills.

How to apply: Cover letter, resume, and writing sample

Solar Energy Industries Association (SEIA)

Address: 122 C Street, NW, 4th Floor
City: Washington   **State:** DC   **Zip:** 20001

Telephone: (202) 383-2600   **Fax:** (202) 383-2670
E-mail: 71263.37@compuserve.com
URL: Not listed

Other locations: None listed

Founded: 1974

Mission:
SEIA is the national trade organization of the photovoltaics, solar thermal manufacturers, and component suppliers responsible for the promotion of solar energy in the US. SEIA represents the interests of the industry to legislators on Capitol Hill, federal agencies and the administration.

Programs:
SEIA co-sponsors and participates in dozens of conferences each year with the Department of Energy, US Agency for International Development, and other organizations. SEIA hosts SOLTECH, an annual conference which attracts over 1,000 people showcasing the latest developments in solar systems for utilities, state and federal facilities, international markets, and residences. SEIA publishes *The Solar Industry Journal,* a quarterly magazine for members. SEIA also puts out a *Sun Flash* report, listing market opportunities in solar energy.

Atmosphere: Casual, often hectic

Volunteers: 0

STAFF POSITIONS
Full-time staff: 19   **Part-time staff:** 0
Staff positions available: Varies
Staff salary: Depends on position

INTERNSHIPS
How Many: Varies   **Duration:** Varies
Remuneration: Unpaid

Duties:
Research, writing, database management, and other tasks depending on skills and abilities.

Qualifications:
Enthusiastic, positive attitude, willing to learn, reliable in doing work, as well as prompt and consistent arrival.

How to apply: Cover letter and resume

Southeast Alaska Conservation Council

Address: 419 Sixth Street, Suite 328
City: Juneau   **State:** AK   **Zip:** 99801

Telephone: (907) 586-6942   **Fax:** (907) 463-3312
E-mail: seacc@igc.apc.org
URL: http://www.juneau.com/seacc/

Other locations: None listed

Founded: 1969

Mission:
SEACC's purpose is to ensure that a substantial portion of Southeast Alaska is retained and protected in a minimally changed condition, while also encouraging human enjoyment and use of these remarkable resources in a manner consistent with retaining them substantially unimpaired. SEACC's goal is to achieve a system of reservations of various sizes and types in a specific management framework that will achieve these objectives.

Programs:
SEACC works through grassroots organizing and education projects involving Southeast Alaskan communities and interested parties throughout the nation. These efforts include road shows, public meetings, and slide shows. Publications include a quarterly newsletter, numerous brochures, *Taking Charge* (a manual to educate people on how to become involved with timber sales), and *Our Forest Home*, which discusses sustainable development.

Atmosphere: Casual, hectic most of the time

Volunteers: 5

STAFF POSITIONS
Full-time staff: 7   **Part-time staff:** 1
Staff positions available: 1 to 2 per year
Staff salary: $25,000 to $28,000 per year plus benefits

INTERNSHIPS
How Many: Varies   **Duration:** Varies
Remuneration: Unpaid

Duties:
Legal research on timber sales, air and water quality standard violations, grassroots organizing, and helping with publications.

Qualifications:
Interest in Alaskan environmental issues, some environmental studies background, forestry studies, and economic analysis. Legal interns must have completed at least one year of law school.

How to apply: Cover letter, resume, and writing sample

Southern Technology Council

Address: 5001 S. Miami Boulevard, PO Box 12293
City: Research Triangle Park **State:** NC **Zip:** 27709

Telephone: (919) 941-5145 **Fax:** (919) 941-5594
E-mail: LGT@ENCORE.NCREN.NET
URL: Not listed

Other locations: None listed

Founded: 1986

Mission:
The Council seeks to strengthen regional economy through more effective development, commercialization, and deployment of technology; fosters cooperative initiatives among science and technology industries, government, and educational institutions; and functions as a forum for information and recommendations on practices, strategies, policies, and programs. The Council serves its constituency in a hands-on, action-oriented, and practical manner.

Programs:
Industrial Modernization and Extension—helps expand the size, reach, interaction, and effectiveness of related programs.
R&D Infrastructure and Technology Transfer—involves increasing the scope and commercialization of regional R&D.
New Enterprise Development—regards the development of the organizational, financial, and policy infrastructure that helps incubate technology-based enterprises.
Work Force Development—relates to the role and needs of leading-edge industry, changes in education and training, and quality of new entrants to the work force.

Atmosphere: Casual, external interaction brings periods of formality

Volunteers: Varies

STAFF POSITIONS
Full-time staff: 4 **Part-time staff:** 7
Staff positions available: Varies
Staff salary: Varies

INTERNSHIPS
How Many: Varies **Duration:** Varies
Remuneration: Depends on experience
Duties:
A wide range of duties corresponding to the intern's background, strengths, interests, as well as organizational needs.
Qualifications:
A background and interest in science and technology policy, commercialization, and economic development.

How to apply: Cover letter and resume

JOBS YOU CAN LIVE WITH 1996

Southface Energy Institute

Address: PO Box 5506
City: Atlanta   **State:** GA   **Zip:** 30307

Telephone: (404) 525-7657   **Fax:** (404) 525-6420
E-mail: steve@southface.org
URL: http://southface.org/~southfac

Other locations: None listed

Founded: 1978

Mission:
The mission of the Southface Energy Institute is to promote energy-efficient ways of life.

Programs:
The major programs of the Southface Energy Institute involve: a home-building school, utility training, assistance for affordable housing groups, public education, demonstration center, and an internship program.

Atmosphere: Casual and very busy

Volunteers: Varies

STAFF POSITIONS
Full-time staff: 7   **Part-time staff:** 3
Staff positions available: 1 to 2 per year
Staff salary: $18,000 per year for non-engineer, $35,000 per year for engineer

INTERNSHIPS
How Many: 16 every year   **Duration:** 3 to 6 months
Remuneration: $250 rent plus $335 stipend
Duties:
Staff support, interaction with public, educational support.

Qualifications:
Bright, a college junior or senior preferred, but all ages accepted.

How to apply: Application; phone calls or e-mail welcome

Southwest Network for Environmental and Economic Justice

Address: PO Box 7399
City: Albuquerque   **State:** NM   **Zip:** 87194

Telephone: (505) 242-0416   **Fax:** (505) 242-5609
E-mail: SNEEJ@igc.org
URL: Not listed

Other locations: None listed

Founded: 1990

Mission:
The Southwest Network of Environmental and Economic Justice brings together activists and grassroots organizations of people of color from across the Southwest and West to promote regional strategies and perspectives on environmental degradation and other social, racial, and economic injustices.

Programs:
Campaigns—areas include: border justice, Environmental Protection Agency accountability, youth leadership development, high-technology, sovereignty and dumping on Native lands.
Southwest Workplace Toxics Training Project—a collaboration with the labor occupational health program at the University of California-Berkeley.

Atmosphere: Casual, hectic

Volunteers: 2

STAFF POSITIONS
Full-time staff: 6   **Part-time staff:** 1
Staff positions available: 2 to 3 per year
Staff salary: $15,000 to $18,000 per year to start

INTERNSHIPS
How Many: Varies   **Duration:** Varies
Remuneration: Varies

Duties:
Administrative and regional office staffing and helping on specific campaigns and projects.

Qualifications:
Interns must be a person of color, have some community student or worker organizing exposure, have a social justice perspective, and Spanish language skills are a plus.

How to apply: Cover letter and resume

JOBS YOU CAN LIVE WITH 1996

Southwest Research and Information Center (SRIC)

Address: PO Box 4524, 105 Stanford, SE
City: Albuquerque   **State:** NM   **Zip:** 87106

Telephone: (505) 262-1862   **Fax:** (505) 262-1864
E-mail: Not listed
URL: Not listed

Other locations: None listed

Founded: 1971

Mission:
Southwest Research and Information Center (SRIC) exists to provide timely, accurate information to the public on matters that affect the environment, human health, and communities in order to protect natural resources, promote citizen participation, and ensure environmental and social justice now and for future generations.

Programs:
SRIC's work can be described in two ways.
Provide services—technical assistance, networking, public information, policy analysis, environmental analysis, and skills development.
Integrate services into programs—Broader Environment; Community Water, Wastes, and Toxics; Environmental Information and Education; Nuclear Waste Safety.
Publications include *The Workbook,* which is published quarterly and used by citizens nationwide.

Atmosphere: Mix of work styles not easily categorized

Volunteers: 0

STAFF POSITIONS
Full-time staff: 10   **Part-time staff:** 0
Staff positions available: 0 to 1 per year
Staff salary: Not listed

INTERNSHIPS
How Many: 3   **Duration:** 3 to 6 months
Remuneration: Unpaid
Duties:
Depending on the program, the student could need legal knowledge, good writing skills, bilingual language skills, and ability to work with minority communities.
Qualifications:
Course experience in the issue area in which the intern wants to work.

How to apply: Cover letter, resume, academic transcript, and writing sample

Structural Dynamics Research Corporation (SDRC)

Address: 2000 Eastman Drive
City: Milford **State:** OH **Zip:** 45150-2789

Telephone: (513) 576-2554 **Fax:** (513) 576-2554
E-mail: brian.wills@sdrc.com
URL: http://www.sdrc.com/

Other locations: Germany, Japan, France, Spain, Italy, India, Korea, China, Switzerland, Sweden, England, and Belgium. In the US, there are offices in CA, VA, MN, AZ, FL, GA, and other states
Founded: 1967

Mission:
The SDRC works to provide automotive, manufacturing, and engineering industries with software to automate their processes. SDRC stresses being "good citizens" as members of the corporate community.

Programs:
Specific programs vary throughout all of the offices, but two publications *Dimension* and *Working Ideas* are the most widely distributed.

Atmosphere: Usually casual

Volunteers: Varies

STAFF POSITIONS
Full-time staff: 1200 in all **Part-time staff:** Varies
Staff positions available: Varies

Staff salary: $35,000 to $40,000 per year on average

INTERNSHIPS
How Many: 30 to 50 per semester **Duration:** Roughly 4 months
Remuneration: Approximately $9.15 per hour
Duties:
Exposure to state-of-the-art software application development and methodologies.

Qualifications:
High GPA, working on BSCS or BSME or a related field.

How to apply: Cover letter, resume, academic transcript, and writing sample

JOBS YOU CAN LIVE WITH 1996

Student Conservation Association—Resource Assistant Program

Address: PO Box 550
City: Charlestown **State:** NH **Zip:** 03603

Telephone: (603) 543-1700 **Fax:** (603) 543-1828
E-mail: Not listed
URL: Not listed

Other locations: Washington, DC; Denver, CO; Oakland, CA; and Los Angeles, CA
Founded: 1957

Mission:
The Student Conservation Association—Resource Assistant Program offers opportunities for education, leadership, and personal development while providing the highest quality public service in conservation.

Programs:
The Resource Assistant Program offers adults 18 and older a choice of 1,200 expense-paid volunteer positions annually. Positions are about 12 weeks long and range widely from hydrology research and management to wilderness search and back country patrol, from endangered species research to natural resource management.

Atmosphere: Atmosphere varies

Volunteers: Varies

STAFF POSITIONS
Full-time staff: 2 to 3   **Part-time staff:** 2 to 3
Staff positions available: Varies
Staff salary: Varies

INTERNSHIPS
How Many: Varies   **Duration:** 12 weeks
Remuneration: Travel, housing, and food
Duties:
Varies widely, but all interns are treated like professional staff and are expected to complete their work to the highest standards.

Qualifications:
Eagerness to learn and the ability to work hard.

How to apply: Application, academic transcript, and 3 references

Student Conservation Association (SCA)

Address: PO Box 550
City: Charlestown **State:** NH **Zip:** 03603

Telephone: (603) 543-1700 **Fax:** (603) 543-1828
E-mail: Not listed
URL: Not listed

Other locations: Denver, CO; Los Angeles, CA; Newark, NJ; Oakland, CA; and Seattle, WA
Founded: 1957

Mission:
The Student Conservation Association fosters lifelong stewardship of the environment by offering opportunities for education, leadership, and personal development while providing the highest quality public service in natural resource management, environmental protection and conservation.

Programs:
SCA coordinates: national high school program, conservation career development program, New Hampshire Conservation Corps, Henry S. Francis, Jr. Wilderness Work Skills Program, SCA/AmeriCorps, and Earth Work Service Days. In addition to these programs, SCA work includes monitoring public policy and political lobbying, community organnizing, and a focus on racial, ethnic, and gender issues.

Atmosphere: Casual, businesslike

Volunteers: 2,000 nationwide

STAFF POSITIONS
Full-time staff: 60 **Part-time staff:** 3
Staff positions available: 6 to 7 per year
Staff salary: Varies

INTERNSHIPS
How Many: Varies **Duration:** 2 to 4 months
Remuneration: Benefits vary
Duties:
Depends on program.
Qualifications:
A strong interest in environmental issues.

How to apply: Cover letter, resume, application, transcript, writing sample, and references

Student Pugwash USA (SPUSA)

Address: 815 15th Street, NW, Suite 814
City: Washington **State:** DC **Zip:** 20005

Telephone: (202) 393-6555 **Fax:** (202) 393-6550
E-mail: spusa@spusa.org
URL: http://www.spusa.org/pugwash/

Other locations: None

Founded: 1979

Mission:
SPUSA promotes the socially responsible application of science and technology. SPUSA encourages young people to examine ethical, social, and global implications of science and technology, and to make these concerns a guiding focus of academic and professional endeavors. SPUSA is based on the 1995 Nobel Peace Prize winners, Pugwash Conferences on Science and World Affairs, and is 1 of over 20 Student and Young Pugwash groups around the world.

Programs:
Pledge Campaign—a promise to work for a better world.
Events and Activities—biennial international conferences that examine cutting-edge issues; national conference, with both leadership skills and issue examination; regional events; chapter activities on over 50 campuses; and mentorship opportunities.
Publications—*Chapter Organizing Guide*; newsletters: *Tough Questions* and *Pugwatch*; *mind•full* issue brief series; *Global Issues Guidebook*, authored by students; *Jobs You Can Live With*.

Atmosphere: Casual, friendly, fast-paced

Volunteers: 0

STAFF POSITIONS
Full-time staff: 4 **Part-time staff:** 0
Staff positions available: 0 to 1 per year
Staff salary: Low twenties to start

INTERNSHIPS
How Many: Varies **Duration:** Semester or summer
Remuneration: $5.25 per hour; increases if over 4 months

Duties:
Interns assist in all phases of producing publications, coordinate conferences and events, develop and maintain programs, and help with various administrative tasks.

Qualifications:
An interest in how science and technology affect society, creativity, detail-oriented, self-motivated, and able to work independently and as part of a team. Other skills depend on project.

How to apply: Cover letter, resume, and brief writing sample

Taking Off

Address: PO Box 104
City: Newton **State:** MA **Zip:** 12161

Telephone: (617) 639-1606 **Fax:** (617) 630-1605
E-mail: None listed
URL: None listed

Other locations: None listed

Founded: 1995

Mission:
Taking Off matches students interested in finding internship opportunities in the US or in other countries with appropriate organizations. These services can assist individuals in finding positions covering their area of interest, including scientific and technological issues.

Programs:
In addition to matching students with organizations, Taking Off also counsels students about the types of opportunities available. For example, students would be asked to describe the issues on which they are passionate, what their plans for the future include, and the values and purposes of further education. This allows Taking Off to meet the needs and goals of both student and potential employer.

Atmosphere: Not Listed

Volunteers: 1

STAFF POSITIONS
Full-time staff: 1 **Part-time staff:** 0
Staff positions available: None
Staff salary: None listed

INTERNSHIPS
How Many: Varies **Duration:** Varies
Remuneration: Varies
Duties:
Varies, depending on organization and project.

Qualifications:
Varies, depending on organization and project.

How to apply: Phone call

JOBS YOU CAN LIVE WITH 1996

Teach for America

Address: 20 Exchange Place, 8th Floor
City: New York **State:** NY **Zip:** 10005

Telephone: (212) 425-9039 **Fax:** (212) 425-9347
E-mail: TFA@aol.com
URL: http://www.wam.umd.edu/~agordon/tfahome.html

Other locations: 13 regional offices in the US

Founded: 1989

Mission:
Teach for America is the national teacher corps. The mission is to build a diverse corps of outstanding recent college graduates who commit two years to teach in under-resourced urban and rural public schools, and who serve as life-long leaders in the pursuit of academic excellence and equity.

Programs:
Teach for America recruits recent college graduates from all academic majors who commit two years to teach. In many school districts, there is an overwhelming need for individuals with a background in mathematics, the sciences, and technology.

Atmosphere: Varies

Volunteers: 0

STAFF POSITIONS
Full-time staff: 75 **Part-time staff:** 0
Staff positions available: Usually none
Staff salary: Varies

INTERNSHIPS
How Many: 12 **Duration:** Summer
Remuneration: $1,000, travel, room, board
Duties:
Logistical and administrative work for the 5-week summer institute.
Qualifications:
Organizational skills, leadership, and initiative.

How to apply: Cover letter, resume, application, academic transcripts, and writing samples

Tellus Institute, Inc.

Address: 11 Arlington Street
City: Boston   **State:** MA   **Zip:** 02116-3411

Telephone: (617) 266-5400   **Fax:** (617) 266-8303
E-mail: POSTMASTER@TELLUS.COM
URL: http://www.channel/1.com/users/tellus/tellus.html

Other locations: None listed

Founded: 1977

Mission:
The Institute conducts policy, regulatory, and technical studies with the aim of advancing environmental, resource, and development strategies that are rational, equitable, and sustainable. The Institute is committed to developing and disseminating innovative analytic methods, to strengthening human and institutional capacity, and to offering vision and direction to governments, the public, and all participants in the transition to sustainable societies.

Programs:
Energy—provides expertise on technical, economic, regulatory, environmental, and policy aspects of energy.
Solid Waste—performs research, planning studies, and computer modeling on issues of waste management, materials use, and related economic and environmental impacts.
Risk Analysis—evaluates environmental, public health, and financial risks of projects; provides policy, regulatory, and technical analysis.
Stockholm Environment Institute Boston Center— performs research, policy evaluations, and field applications concerned with environmentally sound development.

Atmosphere: Casual, serious, demanding
Volunteers: 0

STAFF POSITIONS
Full-time staff: 43   **Part-time staff:** 0
Staff positions available: 1 to 2 per year
Staff salary: $25,000 per year for candidates with a B.S.

INTERNSHIPS
How Many: 1 to 2   **Duration:** Varies
Remuneration: Modest stipend
Duties:
Assistance on research, administrative support, number crunching, phone calls, updating mailing lists.
Qualifications:
Intelligence and an interest in environmental and related issues.
How to apply: Cover letter, resume, and academic transcript

JOBS YOU CAN LIVE WITH 1996

Tennessee Environmental Council

Address: 1700 Hayes Street, Suite 101
City: Nashville   **State:** TN   **Zip:** 37203

Telephone: (615) 321-5075   **Fax:** (615) 321-5082
E-mail: christl@edge.net
URL: http://www.nol.com/~nol/tec.html

Other locations: None listed

Founded: 1970

Mission:
The Tennessee Environmental Council is a nonprofit, statewide coalition of individual and organizational members whose mission is to educate and advocate for the protection of Tennessee's environment and public health.

Programs:
The Tennessee Environmental Council publishes a quarterly newsletter titled *Protect*. Major programs are in the areas of groundwater protection, pollution prevention and reduction, and wetlands protection and restoration.

Atmosphere: Fast-paced, casual

Volunteers: Many

STAFF POSITIONS
Full-time staff: 2   **Part-time staff:** 0
Staff positions available: None
Staff salary: $17,000 per year plus health benefits

INTERNSHIPS
How Many: Varies   **Duration:** Varies
Remuneration: Unpaid
Duties:
Research and writing.

Qualifications:
Good research and writing skills, hard working, intelligent, responsible, self-starter.

How to apply: Cover letter, resume, and writing sample

Texas Environmental Center (TEC)

Address: Box 220 S-300
City: Austin   **State:** TX   **Zip:** 78767-0220

Telephone: (512) 479-6669   **Fax:** (512) 473-4097
E-mail: mfrech@tec.org
URL: http://www.tec.org

Other locations: None listed

Founded: 1991

Mission:
TEC is a nonprofit organization using information technology to extend environmental information, education, and stewardship.

Programs:
TEC currently administers a computer bulletin-board system called the Tributary, which hosts a number of collaborative learning projects. These include the Colorado River Watch program for water quality monitoring and the In Concert Program, which engages children in Energy Conservation learning. TEC also acts as a publisher of *Environmental Context* on the Internet. This includes the *Texas Environmental Almanac*, a 400-page research compendium and *Green Beat*, an electronic magazine profiling exemplary environmental efforts of students.

Atmosphere: Semiformal, friendly

Volunteers: 2

STAFF POSITIONS
Full-time staff: 2   **Part-time staff:** 2
Staff positions available: Usually none
Staff salary: $25,000 to $30,000 per year, health insurance

INTERNSHIPS
How Many: 4 to 5 per year   **Duration:** 1 to 2 semesters
Remuneration: Stipend is available
Duties:
Writing, graphic arts, HTML and Web development, computer technician activities, classroom education, community outreach, research, curriculum development, and creative input in all activities.

Qualifications:
Self-efficiency, motivation, creativity, team-oriented, knowledge of environmental issues, policy and information technologies preferred.

How to apply: Cover letter, resume, academic transcript, and writing sample; e-mail or phone calls welcome

The Progressive Magazine

Address: 409 E. Main Street
City: Madison   **State:** WI   **Zip:** 53703

Telephone: (608) 257-4626   **Fax:** (608) 257-3373
E-mail: progressive@peacenet.org
URL: Not listed

Other locations: None listed

Founded: 1909

Mission:
The Progressive Magazine is a monthly, political magazine focusing on peace and social justice.

Programs:
All programs revolve around the production of the magazine.

Atmosphere: Causal, relaxed

Volunteers: 2

STAFF POSITIONS
Full-time staff: 12   **Part-time staff:** 2
Staff positions available: 0 to 1 per year
Staff salary: $19,000 per year, full benefits, 12 vacation days a year

INTERNSHIPS
How Many: 1   **Duration:** Semester or summer
Remuneration: Unpaid
Duties:
Editorial and general support.
Qualifications:
Journalism experience and political commitment.

How to apply: Cover letter, resume, and writing sample

Transportation Alternatives

Address: 82 St. Marks Place
City: New York   **State:** NY   **Zip:** 10009

Telephone: (212) 475-4800   **Fax:** (212) 475-4551
E-mail: TRANSALT@ECHO.COM
URL: Not listed

Other locations: None listed

Founded: 1973

Mission:
Transportation Alternatives promotes walking, cycling, and better mass transit in and around New York City.

Programs:
Transportation Alternatives is involved with two major programs: the Bicycle Program and the Pedestrian Program. Both programs employ at least 1 full-time advocate to work at both grassroots and policy-planning levels to address concerns specific to these programs. Transportation Alternatives also produces two publications: *Transportation Magazine*, which encompasses all campaigns, and *City Cyclist*, which specifically deals with cyclist issues.

Atmosphere: Casual, very dedicated

Volunteers: Many

STAFF POSITIONS
Full-time staff: 5   **Part-time staff:** 2
Staff positions available: None
Staff salary: Varies

INTERNSHIPS
How Many: Varies   **Duration:** Varies
Remuneration: Unpaid
Duties:
Varies.

Qualifications:
Interest in related issues and dedication to the cause.

How to apply: Cover letter

Traprock Peace Center

Address: 103 A Keets Road, Woolman Hill
City: Deerfield **State:** MA **Zip:** 01342

Telephone: (413) 773-7427 **Fax:** (413) 773-7427
E-mail: Not listed
URL: Not listed

Other locations: None listed

Founded: 1979

Mission:
The Traprock Peace Center seeks to teach nonviolence, promote a sustainable economy, build community, help to end war, and take leadership on environmental issues.

Programs:
The Center publishes a *Peace Action Report* that is distributed about four times a year. Recent events include: a gathering to further the conversations with women who are "Back from Beijing," celebrations of Gandhi's and Elizabeth Cady Stanton's birthdays, and a lecture series on the roots of violence. All are designed to promote discourse and change.

Atmosphere: Hectic moments, relaxed as possible

Volunteers: 15

STAFF POSITIONS
Full-time staff: 1 **Part-time staff:** 0
Staff positions available: None
Staff salary: $12 per hour plus benefits

INTERNSHIPS
How Many: Varies **Duration:** 1 to 3 months
Remuneration: Possibility of housing
Duties:
Depends on specific project.

Qualifications:
Thoughtful, kind, persistent, communicative, committed, and ingenious.

How to apply: Cover letter, resume, and writing sample; phone calls welcome

JOBS YOU CAN LIVE WITH 1996

Trees for the Future

Address: 11306 Estonia Drive, PO Box 1786
City: Silver Spring   **State:** MD   **Zip:** 20915

Telephone: (800) 643-0001   **Fax:** (301) 929-0439
E-mail: Not listed
URL: Not listed

Other locations: None listed

Founded: 1989

Mission:
The mission for Trees for the Future is to initiate tree planting at the grassroots level and to develop projects that bring environmental benefit while improving the quality of life for involved communities. This program further informs the American public about the consequences of on-going environmental destruction.

Programs:
Areas of interest include: environmental education and conservation, alley-cropping, erosion control, nursery establishment, and tree planting. Trees for the Future also produces the *Johnny Ipil-Seed Newsletter*.

Atmosphere: Casual, sometimes hectic

Volunteers: 6

STAFF POSITIONS
Full-time staff: 7   **Part-time staff:** 2
Staff positions available: 1 per year
Staff salary: Varies

INTERNSHIPS
How Many: Varies   **Duration:** Varies
Remuneration: Varies
Duties:
Depends on intern interest.
Qualifications:
Appropriate natural resources degree and overseas experience.

How to apply: Cover letter, resume, application, academic transcript, and writing sample; phone calls welcome

JOBS YOU CAN LIVE WITH 1996

Union of Concerned Scientists (UCS)

Address: Two Brattle Square
City: Cambridge   **State:** MA   **Zip:** 02238-9105

Telephone: (617) 547-5552   **Fax:** (617) 864-9405
E-mail: ucs@ucsusa.org
URL: http://www.ucsusa.org

Other locations: Washington, DC and Berkeley, CA

Founded: 1969

Mission:
The Union of Concerned Scientists is dedicated to advancing responsible public policies in areas where science and technology play a critical role. Established in 1969, UCS has created a unique alliance between many of the nation's leading scientists and thousands of committed citizens.

Programs:
UCS conducts technical studies and public education and seeks to influence government policy at the local, state, federal, and international levels. Programs revolve around global resources program, arms control and international security, and transportation and energy.

Atmosphere: Casual, hectic at times

Volunteers: 5

STAFF POSITIONS
Full-time staff: 30   **Part-time staff:** 2
Staff positions available: Varies
Staff salary: Varies greatly

INTERNSHIPS
How Many: 1 to 2 per office   **Duration:** Varies
Remuneration: $1,000 to $1,500 per month
Duties:
Depends on specific program.

Qualifications:
Welcome applicants with education, training, or experience in a wide variety of fields.

How to apply: Cover letter, resume, application, two references, and writing sample

University of Rhode Island Environmental Education Center

Address: W. Alton Jones Campus, URI, 401 Victory Highway
City: W. Greenwich   **State:** RI   **Zip:** 02817

Telephone: (401) 397-3304 x6043   **Fax:** (401) 397-3293
E-mail: shun2741@URIACC.URI.EDU
URL: Not listed

Other locations: None listed

Founded: 1962

Mission:
The Center's mission is to provide exciting learning experiences for youth in the natural environment, which leads to a heightened sense of responsibility towards improving the quality of all life on our planet.

Programs:
Programs include: overnight environmental summer camp, summer teen expeditions, Farm and Forest Summer Day Camp, residential environmental school field trips, farm day field trips for schools, and other public programs.

Atmosphere: Casual, relaxed, hectic

Volunteers: 30

STAFF POSITIONS
Full-time staff: 4   **Part-time staff:** 3
Staff positions available: 8 seasonal positions
Staff salary: Varies

INTERNSHIPS
How Many: 4   **Duration:** 4 to 6 months
Remuneration: Room, board, $50 per week

Duties:
All are teaching internships—working with children between 5 and 14 years of age. Some light maintenance is required.

Qualifications:
Enthusiasm and a demonstrated dedication to the environment, children, and education.

How to apply: Cover letter and resume

Vermont Raptor Center

Address: RR2 Box 532
City: Woodstock **State:** VT **Zip:** 05091

Telephone: (802) 457-2779 **Fax:** (802) 457-4861
E-mail: Not listed
URL: Not listed

Other locations: None listed

Founded: 1987

Mission:
The Vermont Raptor Center promotes avian conservation on a personal level through education, raptor rehabilitation, and research.

Programs:
Living Museum—houses 25 species of nonreleaseable raptors in large flight habitats on year-round display.
Rehabilitation—cares for more than 600 injured birds yearly in an effort to return them to their native habitat. The most modern techniques are used to provide the highest quality of care.
Education—focuses on environmental education. Programs create an awareness about raptors, their conservation, and the role they play in the environment.
Publications—distributed by the Center on raptor related issues.

Atmosphere: Casual, hectic at times

Volunteers: 150

STAFF POSITIONS
Full-time staff: 2 **Part-time staff:** 2
Staff positions available: 0 to 1 per year
Staff salary: $18,000 to $22,000 per year, health benefits

INTERNSHIPS
How Many: 3 per session **Duration:** 3 to 5 months
Remuneration: Housing, $750/month
Duties:
90% of the intern's time is spent in raptor rehabilitation and 10% in environmental education.
Qualifications:
Enthusiasm, flexibility, a strong interest in birds, 18 years or older, and the ability to lift up to 50 pounds of materials and equipment.

How to apply: Cover letter, resume, and two letters of reference

Video Project

Address: 5332 College Avenue
City: Oakland **State:** CA **Zip:** 94618

Telephone: (510) 655-9050 **Fax:** (510) 655-9115
E-mail: videoproject@videoproject.org
URL: http://www.videoproject.org/videoproject/

Other locations: None listed

Founded: 1983

Mission:
Founded by Oscar-winning documentary filmmaker Vivienne Verdon-Roe and Oscar-nominee Ian Thiermann, the Video Project is the nation's only nonprofit organization specializing in distribution of videos on environmental issues and related global concerns. The catalog of more than 200 programs from over 150 filmmakers is one of the largest collections of environmental videos in the US and includes some of the finest programs on these issues available.

Programs:
The major programs involve practical how-to series on various environmental issues including cleaning up toxics and preserving ancient forests. The Video Project also helps distribute videos from other environmental organizations including Greenpeace. These videos are an essential tool to provide vital information the public might not be receiving from the mass media, and to promote these programs at an affordable price so they can have the greatest impact to the broadest audience.

Atmosphere: Casual, relaxed, spirited

Volunteers: Varies

STAFF POSITIONS
Full-time staff: 6 **Part-time staff:** 0
Staff positions available: None
Staff salary: Varies

INTERNSHIPS
How Many: Varies **Duration:** Varies
Remuneration: Unpaid
Duties:
Provide marketing assistance, write press releases, help send videos out for review, establish World Wide Web links, and coordinate mailings.

Qualifications:
Computer, interpersonal, and office skills.

How to apply: Cover letter and resume; phone calls welcome

Visions in Action

Address: 3637 Fulton Street, NW
City: Washington **State:** DC **Zip:** 20007

Telephone: (202) 625-7403 **Fax:** (202) 625-2353
E-mail: AFRICAJOB@aol.com
URL: Not listed

Other locations: Kenya, Uganda, Tanzania, Zimbabwe, South Africa, Mexico, Dominican Republic, and Burkina Faso
Founded: 1988

Mission:
Visions in Action provides volunteers to development organizations in Africa and Latin America. Volunteers live in a supportive community, following a grassroots approach to development. Positions are available for one year or six months.

Programs:
Volunteer opportunities are available in; agriculture, appropriate technology, child development, communications and journalism, education, environment and natural resources, food and nutrition, family planning, health, housing, human rights and law, mental disabilities, physical disabilities, natural science, refugees and relief, small business, social science, women, and youth.

Atmosphere: Casual, hectic, fun

Volunteers: 80

STAFF POSITIONS
Full-time staff: 8 **Part-time staff:** 6
Staff positions available: 1 to 2 per year
Staff salary: Varies

INTERNSHIPS
How Many: Unlimited **Duration:** 3 months
Remuneration: Unpaid
Duties:
Fundraising, public relations, media, administration, and campus recruitment.

Qualifications:
Hard working, flexible, willing to do whatever it takes.

How to apply: Cover letter and resume

Volunteers for Outdoor Colorado

Address: 600 S. Marion Parkway
City: Denver **State:** CO **Zip:** 80209

Telephone: (303) 715-1010 **Fax:** (303) 715-1212
E-mail: Not listed
URL: Not listed

Other locations: None listed

Founded: 1984

Mission:
Volunteers for Outdoor Colorado is a nonprofit organization with a mission to instill in Coloradans a sense of personal responsibility for their public lands, thereby creating active and beneficial stewardship of the state's precious resources.

Programs:
Volunteer for Outdoor Colorado—12 trail-building and tree-planting projects that involve volunteer labor and the surrounding communities. Most work is performed on public lands.
Volunteer for Outdoor Colorado Clearinghouse—matches willing volunteers with federal, state, and local land management agencies who are in need of volunteers.
Trail Breaking News—a newsletter for members.

Atmosphere: Casual, hectic

Volunteers: 2,500

STAFF POSITIONS
Full-time staff: 6 **Part-time staff:** 2
Staff positions available: 1
Staff salary: $18,000 to $23,000 per year, must have college degree

INTERNSHIPS
How Many: 1 to 3 **Duration:** Flexible
Remuneration: Negotiable

Duties:
Writing and editing newspaper articles, developing volunteer resources, organizing the management of National Trails Day, and countless other tasks.

Qualifications:
Committed to environmental issues, at least junior standing in college, own transportation, computer experience, team player, and physical stamina.

How to apply: Cover letter, resume, and writing sample; phone calls welcome

War and Peace Foundation

Address: 32 Union Square East
City: New York **State:** NY **Zip:** 10003

Telephone: (212) 777-4210 **Fax:** (212) 995-9652
E-mail: Not listed
URL: Not listed

Other locations: None listed

Founded: 1981

Mission:
The mission of The War and Peace Foundation has three major areas of focus:
1. To rid the world of nuclear weapons and nuclear reactors.
2. To promote adequate funding to the United Nations.
3. To use the media to further these purposes.

Programs:
Programs usually revolve around current related events. Recent focus includes the World Court's decision to ban nuclear weapons. The Foundation also publishes the *War and Peace Digest*, covers issues on cable television, and produces videos on relevant topics.

Atmosphere: Casual, hectic

Volunteers: 0

STAFF POSITIONS
Full-time staff: 3 **Part-time staff:** 2
Staff positions available: None
Staff salary: Not listed

INTERNSHIPS
How Many: 1 to 2 **Duration:** Varies
Remuneration: $5 per hour
Duties:
Administrative duties and research.

Qualifications:
Sympathy and interest in related issues with office, computer skills a plus.

How to apply: Phone call

War Resisters League (WRL)

Address: 339 Lafayette Street
City: New York   **State:** NY   **Zip:** 10012

Telephone: (212) 228-0450   **Fax:** (212) 228-6193
E-mail: wrl@igc.apc.org
URL: Not listed

Other locations: Norwich, CT

Founded: 1923

Mission:
Believing war to be a crime against humanity, WRL advocates Gandhian nonviolence as the method for creating a democratic society free of war, racism, sexism, and human exploitation.

Programs:
WRL produces *The Nonviolent Activist*, a 24-page bimonthly publication that focuses on ongoing disarmament work. One continuing program is Youth Peace, which promotes nonviolence, justice, and an end to the militarization of youths.

Atmosphere: Casual, busy, hectic at times

Volunteers: 4

STAFF POSITIONS
Full-time staff: 2   **Part-time staff:** 2
Staff positions available: 1 per year at most
Staff salary: $25,000 per year and full benefits

INTERNSHIPS
How Many: Varies   **Duration:** Varies
Remuneration: Unpaid
Duties:
Program and office tasks, and special projects.

Qualifications:
Enthusiasm and a strong interest in pacifism.

How to apply: Cover letter and resume

Washington Wilderness Coalition

Address: 4649 Sunnyside Avenue N.
City: Seattle   **State:** WA   **Zip:** 98103

Telephone: (206) 633-1992   **Fax:** (206) 633-1996
E-mail: WAWILD@AOL.COM
URL: Not listed

Other locations: None listed

Founded: 1979

Mission:
WWC coordinates volunteers and activists to meet with government officials and develops conservation strategies. As an educational organization, WWC provides technical information and assistance on conservation issues to individuals, members, and member organizations. WWC has created a vital, growing network of people throughout the state who are dedicated to protecting and preserving Washington's heritage of wild lands.

Programs:
Roadless Area Monitoring and Mapping—mapping project to survey all roadless systems in the state.
Roadless Area Protection Team—grassroots activist network mobilizing on current issues and coordinating speakers and educational programs.
Wilderness Education—Something Aboard program has taught over 50,000 Scouts (boys and girls) to employ low-impact camping techniques.

Atmosphere: Casual, hectic, fun, productive

Volunteers: 15

STAFF POSITIONS
Full-time staff: 4   **Part-time staff:** 1
Staff positions available: 1 per year
Staff salary: $18,000 per year to start

INTERNSHIPS
How Many: 10   **Duration:** 3 months to 1 year
Remuneration: Unpaid
Duties:
Varies depending on participant interest.
Qualifications:
An interest in environmental issues.

How to apply: Phone call

Wildlife Center of Virginia

Address: PO Box 1557
City: Waynesboro   **State:** VA   **Zip:** 22980

Telephone: (540) 942-9453   **Fax:** (540) 943-9453
E-mail: Not listed
URL: Not listed

Other locations: None listed

Founded: 1982

Mission:
The Wildlife Center of Virginia, a hospital for native wild animals, promotes positive values and behaviors toward wildlife and the environment through education, research and veterinary care. The Center provides vision leadership training and support to empower individuals, professionals, and institutions to care for and protect wildlife and the environment.

Programs:
Programs involve a comprehensive research and teaching hospital for native wildlife and environmental education programs for student and adult groups. Publications include an annual report, *Wildlife Center Teacher's Project*, and *Handbook of Wildlife Medicine*.

Atmosphere: Casual, hectic at times

Volunteers: 150

STAFF POSITIONS
Full-time staff: 10   **Part-time staff:** 8
Staff positions available: 2 to 3 per year
Staff salary: Varies

INTERNSHIPS
How Many: 3   **Duration:** 1 to 3 months
Remuneration: Free or low-cost housing
Duties:
Three types of internships are available: animal caretaker, communication and marketing, and environmental education. Duties vary between the different internships.
Qualifications:
Prior experience in the internship area, a minimum 2 years college, and strong communication skills.
How to apply: Cover letter, resume, and 2 references

Wildlife Conservation Society

Address: 2300 Southern Boulevard
City: Bronx   **State:** NY   **Zip:** 10460

Telephone: (718) 220-5087   **Fax:** (718) 220-2464
E-mail: Not listed
URL: Not listed

Other locations: New York, NY; Brooklyn, NY; and Queens, NY

Founded: 1895

Mission:
Wildlife Conservation Society works to preserve wild animals in the wilderness and to promote animal conservation and education.

Programs:
The society runs the largest urban zoo in the country and also operates 3 other zoos and an aquarium in the New York area. The society sponsors hundreds of fellowships each year, is responsible for hundreds of field conservation projects, and publishes a magazine on a bimonthly basis.

Atmosphere: Casual to formal

Volunteers: 500

STAFF POSITIONS
Full-time staff: 800   **Part-time staff:** 300
Staff positions available: 15 per year

Staff salary: $25,000 to $30,000 per year

INTERNSHIPS
How Many: Varies   **Duration:** Varies
Remuneration: Small stipend available
Duties:
Teaching environmental science to a K-12 audience.

Qualifications:
Exemplary academic background, and environmental and animal work or experience.

How to apply: Cover letter and resume

William Penn House

Address: 515 E. Capitol Street, SE
City: Washington   **State:** DC   **Zip:** 20003

Telephone: (202) 543-5560   **Fax:** (202) 543-3814
E-mail: dirpennhouse@igc.apc.org
URL: http://www.quaker.org

Other locations: None listed

Founded: 1966

Mission:
William Penn House provides seminar programs to interested groups from schools, colleges and universities, churches, and other organizations. The topics are chosen by the group and are in all areas of peace and social justice concerns. William Penn House also provides overnight accommodations to groups and individuals coming to Washington for activities related to peace and social justice issues.

Programs:
Seminar programs are planned for groups requesting them: from time to time they hold training programs, workshops, and panel discussions. *Penn Notes* is their quarterly newsletter.

Atmosphere: Busy, informal

Volunteers: 0

STAFF POSITIONS
Full-time staff: 3   **Part-time staff:** 3
Staff positions available: Varies
Staff salary: Varies

INTERNSHIPS
How Many: 2   **Duration:** 12 months
Remuneration: $3,360 plus room and board
Duties:
Office work, hospitality functions, program work, and interns are required to spend two days at another Washington Peace organization.

Qualifications:
College graduate, flexibility, hard worker, people skills, and sympathy with peace concerns.

How to apply: Cover letter and resume

Women in International Security (WIIS)

Address: CISSM, School of Public Affairs, University of Maryland
City: College Park **State:** MD **Zip:** 20742

Telephone: (301) 405-7612 **Fax:** (301) 403-8107
E-mail: wiis@puafmail.umd.edu
URL: http://www.puaf.umd.edu/wiis

Other locations: None listed

Founded: 1987

Mission:
WIIS is dedicated to enhancing opportunities for women working in foreign and defense policy. An international, nonprofit, nonpartisan network and educational program, WIIS is open to women and men at all stages of their careers. It aims to inform others about roles and contributions women are making in the international security community.

Programs:
WIIS Words Newsletter—exchanges information and highlights the achievements of women.
WIIS Summer Symposium—encourages graduate students to pursue careers in international security studies by introducing them to women specialists in the field.
WIIS Databank—international database of women specialists.
WIIS Seminars and Conferences—professionals in substantive discussions.
Fellowship Guide, Internship Directory, and Jobs Hotlines offer information on opportunities available.

Atmosphere: Fast-paced, pleasant, academic

Volunteers: 35 to 50

STAFF POSITIONS
Full-time staff: 3 **Part-time staff:** 0
Staff positions available: 0 to 1 per year
Staff salary: Depends on education and experience

INTERNSHIPS
How Many: 1 to 2 **Duration:** Semester summer
Remuneration: Small stipend in summer
Duties:
Maintain and update computerized databank, general office duties, conference preparation and research for WIIS publications, assist with special projects.

Qualifications:
Good oral and written communication skills; experience with computers; interest in international relations, foreign policy, or security; and interest in the advancement of women in these fields.

How to apply: Cover letter and resume; fax or e-mail welcome

Women Strike for Peace

Address: 110 Maryland Avenue, NE, #302
City: Washington **State:** DC **Zip:** 20002

Telephone: (202) 543-2660 **Fax:** (202) 544-1187
E-mail: Not listed
URL: Not listed

Other locations: Berkeley, CA; Seattle, WA; and New York, NY

Founded: 1961

Mission:
Women Strike for Peace is an activist movement of primarily volunteer women working to halt militarism. Major priorities are nuclear, chemical, biological, and conventional arms control and disarmament; US intervention in developing countries; and related foreign and domestic policy matters.

Programs:
Women Strike for Peace publishes the *Legislative Alert* newsletter 10 times per year, and their major program is titled Lobby by Proxy.

Atmosphere: Casual, relaxed, sometimes hectic

Volunteers: 5

STAFF POSITIONS
Full-time staff: 0 **Part-time staff:** 0
Staff positions available: None
Staff salary: None

INTERNSHIPS
How Many: 2 to 3 **Duration:** Semester or 3 months
Remuneration: Unpaid
Duties:
Writing and mailing the newsletter, lobbying, and attending congressional briefings and coalition meetings.
Qualifications:
Interest in foreign policy and international affairs, and some computer and writing skills.
How to apply: Cover letter, resume and writing sample; phone calls welcome

Women's Action for New Directions

Address: 691 Massachusetts Avenue
City: Arlington **State:** MA **Zip:** 02174

Telephone: (617) 643-6740 **Fax:** (617) 643-6744
E-mail: Wand@world.std.com
URL: Not listed

Other locations: Washington, DC

Founded: 1982

Mission:
Women's Action for New Directions empowers women to act politically to reduce violence and militarism and redirect military resources toward human and environmental needs.

Programs:
The Women Legislator's Lobby (WiLL) is the only national bipartisan network of women state legislators working to influence federal laws, policies, and budget priorities. Since its inception in 1990, WiLL has grown to include women legislators in all 50 states, representing millions of constituents. One in three women state legislators is a WiLL member. WiLL members are leaders for today and for the future—one in four plan to run for higher office in the next five years. The organization also produces a quarterly bulletin that discusses similar issues.

Atmosphere: Casual

Volunteers: 2

STAFF POSITIONS
Full-time staff: 6 **Part-time staff:** 0
Staff positions available: 1 per year
Staff salary: Varies

INTERNSHIPS
How Many: 2 **Duration:** 3 months
Remuneration: Unpaid
Duties:
Varies

Qualifications:
Varies

How to apply: Cover letter and resume

World Federalist Association

Address: 418 7th Street, SE
City: Washington   **State:** DC   **Zip:** 20003

Telephone: (202) 546-3950   **Fax:** (202) 546-3749
E-mail: WFA@IGC.APC.ORG
URL: http://www.getnet.com/wfa

Other locations: New York, NY; Pittsburgh, PA; and Boston, MA

Founded: 1947

Mission:
The WFA is a nonprofit citizen's organization working to transform the UN into a democratic world federation capable of ensuring peace, economic progress, and environmental protection. The essence of WFA's work is to invest legal and political authority in world institutions in order to deal with problems that can only be treated at the global level, while affirming the sovereignty of the nation-state in matters that are essentially internal.

Programs:
The WFA, a public education organization, has over 70 active chapters nationwide. Programs are focused at the grassroots and the policy-making levels. WFA invites all to join them in the important work of raising the awareness of the American people—and the people of all countries—about the urgent need to restructure and strengthen the United Nations. Programs include: The International Criminal Court Education and Advocacy Program, The United Nations Preventive Diplomacy Project, and the grassroots activist network, Partners for Global Change.

Atmosphere: Casual, hectic

Volunteers: 5

STAFF POSITIONS
Full-time staff: 13   **Part-time staff:** 2
Staff positions available: 1 per year

Staff salary: $20,000 per year, health insurance for recent college graduates

INTERNSHIPS
How Many: 5 per session   **Duration:** 3 months to 1 year
Remuneration: $5.00 per day stipend

Duties:
The interns are expected to make substantial contributions to the work of the organization, provided they are able to demonstrate abilities to match these high standards.

Qualifications:
Strong written and oral communication skills, ability to handle a number of projects simultaneously, self-motivation, commitment to building a better, more peaceful world.

How to apply: Cover letter, resume, and 3-to 5-page writing sample

JOBS YOU CAN LIVE WITH 1996

World Game Institute

Address: 3215 Race Street
City: Philadelphia   **State:** PA   **Zip:** 19104

Telephone: (215) 387-0220   **Fax:** (215) 387-3009
E-mail: Not listed
URL: http://www.worldgame.org/~wgi/

Other locations: Washington State

Founded: 1972

Mission:
Nonprofit and UN sanctioned, the World Game Institute develops solutions to world problems by making the world's vital statistics available to everyone. The mission of the World Game Institute is to foster responsible citizenship on a global society through education, information, and workshop programs.

Programs:
The World Game Institute workshops offer groups of 15 to 200 people an opportunity to learn about and experience the dynamics that effect lives in every region of the world. Special programs are available on cultural diversity, world issues, environment, and gender issues. The Institute also provides lecture programs and software, and researches and publishes activities.

Atmosphere: Casual, sometimes hectic, always busy

Volunteers: 10

STAFF POSITIONS
Full-time staff: 12   **Part-time staff:** 2
Staff positions available: Many every year
Staff salary: $15,000 to $35,000 per year

INTERNSHIPS
How Many: Several   **Duration:** 3 months to 1 year
Remuneration: Unpaid
Duties:
Research, assisting in production and preparation of materials, collection of data, editing, writing, and assisting with grant writing.

Qualifications:
Innovative, ability to follow their own path.

How to apply: Cover letter, resume, and writing sample; phone calls welcome

World Service Authority

Address: 1012 14th Street, NW, Suite 1106
City: Washington **State:** DC **Zip:** 20005

Telephone: (202) 638-2662 **Fax:** (202) 638-0638
E-mail: 76507.2343@compuserve.com
URL: http://www.together.org/orgs/wcw/

Other locations: Tokyo, Japan

Founded: 1954

Mission:
WSA implements human rights laws as defined by UN resolutions and treaties and by customary world law, including the human rights of ethnic and social minorities, and ecological rights. WSA is effecting peace, justice, and ecological and human security by promoting and enforcing world law, and grassroots, democratic constitutional processes at the global level. WSA registers world citizens and issues global identification documents.

Programs:
World Syntegrity Project—arranges group meetings all over the world to involve registered citizens in a constitutional process to develop the structures and institutions that will govern at the world level and that will enforce human rights laws.
Client Correspondence—letters to world citizens and world document holders to resolve informational problems and to explain the legal validity and global recognition of these documents.
Client Cases—assist those whose human rights are violated by officials who illegally detained clients or confiscated documents.
Publication—a newsletter, *World Citizen News*.

Atmosphere: Casual, hectic

Volunteers: 15

STAFF POSITIONS
Full-time staff: 12 **Part-time staff:** 2
Staff positions available: 0 to 1 per year
Staff salary: $20,000 to $25,000 per year plus benefits

INTERNSHIPS
How Many: 6 to 8 per year **Duration:** Minimum of 3 months
Remuneration: Varies
Duties:
Letter writing, administrative tasks, e-mail response, research, and attending panels and conferences.

Qualifications:
One year of college, computer literate, language skills other then English, and multi-cultural and diversity experiences.

How to apply: Cover letter, resume, and writing sample in another language if possible.

JOBS
YOU CAN
LIVE WITH
1996

Worldwatch Institute

Address: 1776 Massachusetts Avenue, NW
City: Washington   **State:** DC   **Zip:** 20036-1904

Telephone: (202) 452-1999   **Fax:** (202) 296-7365
E-mail: worldwatch@igc.apc.org
URL: Not listed

Other locations: None listed

Founded: 1974

Mission:
Worldwatch Institute is a nonprofit, public-interest research institute concentrating on global environmental and environment-related issues. The Institute was founded to raise public awareness of these global threats to such a level that it will foster and support an effective public policy response.

Programs:
The majority of the programs at Worldwatch Institute revolve around global-level integrative and interdisciplinary environmental research. The Institute serves as a vital source for environmental policy analysis and information, with works translated into 27 languages for worldwide distribution. The Institute has several publications: the *Worldwatch Papers*, a series of monographs typically numbering five to six per year; the *State of the World* report, which is published annually; *World Watch*, a bimonthly magazine; the *Environmental Alert* series of books (about 2 a year) on specific topics; and *Vital Signs*, a series of the trends that are shaping our future.

Atmosphere: Not listed

Volunteers: 0

STAFF POSITIONS
Full-time staff: 28   **Part-time staff:** 2
Staff positions available: 1 to 5 per year
Staff salary: Varies

INTERNSHIPS
How Many: Varies   **Duration:** Varies
Remuneration: Unpaid
Duties:
Assisting with all aspects of research.
Qualifications:
Strong research skills and an interest in environment-related issues.

How to apply: Cover letter and resume; phone calls welcome

YES! (Youth for Environmental Sanity)

Address: 706 Frederick Street
City: Santa Cruz **State:** CA **Zip:** 95062

Telephone: (408) 459-9344 **Fax:** (408) 458-0344
E-mail: YES@cruzio.com
URL: Not listed

Other locations: None listed

Founded: 1990

Mission:
YES! inspires young people to make positive environmental changes for the future of life on Earth.

Programs:
YES! has a professional performance troupe which goes to junior high and high schools throughout the United States and Canada, educating young people through dance, music, and theater about the state of the environment and inspiring positive change. YES! facilitates workshops, summer camps, and arranges speakers at major conferences. YES! also produces a newsletter and a book called *Choices for Our Future*.

Atmosphere: Fun, casual, hard working

Volunteers: 0

STAFF POSITIONS
Full-time staff: 5   **Part-time staff:** 2
Staff positions available: 1 to 2 per year
Staff salary: $7 to $8 per hour

INTERNSHIPS
How Many: 1 to 2 per semester   **Duration:** Varies
Remuneration: Unpaid
Duties:
Varies, depending or specific projects.

Qualifications:
Leadership ability, efficiency, and a desire to help make a change.

How to apply: Phone call

Zero Population Growth (ZPG)

Address: 1400 16th Street, NW, Suite 320
City: Washington **State:** DC **Zip:** 20036

Telephone: (202) 332-2200 **Fax:** (202) 332-2302
E-mail: ZPG@igc.apc.org
URL: http://www.zpg.org/zpg

Other locations: None listed

Founded: 1968

Mission:
ZPG is a national, nonprofit organization working to slow population growth and achieve a sustainable balance between the earth's people and its resources. It seeks to protect the environment and ensure a high quality of life for present and future generations. ZPG's education and advocacy programs aim to influence public policies, attitudes, and behavior on national and global population issues and related concerns.

Programs:
ZPG focuses on educating Americans about the global issue of overpopulation, both the number of people and consumption patterns. The grade K-12 education program trains 4,000 teachers annually to use population activities and facts to increase awareness of overpopulation. Chapters and activist programs bring population perspectives into focus locally. Lobbying efforts and letter-writing campaigns encourage political leaders to adopt responsible population and environmental policies. The campus organizing newsletter reaches over 700 organizers on over 300 campuses; the membership newsletter goes to 50,000 members.

Atmosphere: Casual, professional

Volunteers: 10

STAFF POSITIONS
Full-time staff: 25 **Part-time staff:** 5
Staff positions available: 2 to 3 per year
Staff salary: $17,000 to $21,000 per year

INTERNSHIPS
How Many: 3 paid interns **Duration:** 6 months
Remuneration: $1190 per month
Duties:
Depending on area, interns help with research and writing, grassroots organizing, public presentations, and attending coalition meetings. Interns spend about 20% of their time on administrative work.

Qualifications:
Flexible, interest in population issues, support of women's rights, pro-choice, and strong writing skills.

How to apply: Cover letter, resume, writing samples, and references

Section 3
WHERE ELSE CAN I LOOK?

Hopefully, you found organizations that look like they would fit in well with your interests. In addition to those organizations, you may want to check out opportunities to work in the government or in another country. Alternatively, you may have a particular business already in mind and want to see how socially responsible they are.

This section refers you to other sources of information to answer those types of questions. We have also included a list of additional resources that might help you find your niche.

Opportunities with the Government

There are many governmental agencies and offices that explore various aspects of science, technology, and society. Working in federal, state, or local government can provide interesting experiences and perspectives on substantive issues and policy making. The federal government is one of the largest employers in the United States, hiring people in almost every occupational field and filling hundreds of thousands of positions annually. Along with full-time positions, there are internships available with government agencies that work on scientific and technological issues.

There are also internships available in the Capitol Hill (and local) offices of representatives and senators that allow students and young professionals to get a first-hand look at how our government functions. Members of Congress tend to hire interns from their home state. Send a cover letter and resume at least six months prior to your desired starting date. For help locating a specific congressional office, call the Capitol Hill switchboard at (202) 224-3121.

State and local government internships also provide opportunities for unique and valuable experience. If you are interested in legislative internships on the state level, contact your governor's office and both houses of the state legislature. States have departments or agencies that deal with: education, agriculture, natural resources, environment, water, pollution, conservation, energy, public health, and urban development. Most local telephone books have a "Blue Pages" section that lists specific departments of state and local government. Opportunities on the city or county level vary greatly, but contacting your mayor or county executive, along with the city or county council is a great way to begin your search.

Federal Government Agencies Related to Science and Technology

Listed below are some federal government agencies that work on scientific, technological, and societal issues. Contact information for the national offices of these agencies is listed below. Although most are located in the Washington, DC area, many agencies have field offices around the country. Contact them to find out about internship local opportunities or employment.

Arms Control and Disarmament Agency
320 21st Street, NW
Washington, DC 20451
(202) 647-2034

National Institutes of Health
9000 Rockville Pike, #344 Building 1
Bethesda, MD 20892
(301) 496-4461

Acquisition and Technology
3000 Defense Pentagon, 3D-1020
Washington, DC 20301
(703) 695-4893

Department of Education
400 Maryland Avenue, SW
Washington, DC
(202) 401-2000

Department of Energy
1000 Independence Avenue, SW
Washington, DC 20585
(202) 586-5000

Department of the Interior
1849 C Street, NW
Washington, DC 20240
(202) 208-3171

Department of State
2201 C Street, NW
Washington, DC 20520
(202) 647-4000

Environmental Protection Agency
401 M Street, SW
Washington, DC 20460
(202) 260-2090

Federal Communications Commission
1919 M Street, NW
Washington, DC 20554
(202) 418-0200

Federal Energy Regulatory Commission
825 North Capitol Street, NE
Washington, DC 20426
(202) 219-2990

Federal Maritime Commission
800 North Capitol Street, NW
Washington, DC 20573
(202) 523-5773

Food and Drug Administration
5600 Fishers Lane
Rockville, MD 20857
(301) 443-3326

National Aeronautics and Space Administration
300 E Street, SW
Washington, DC 20546
(202) 358-2344

National Institute of Standards and Technology
1 Bureau Drive
Building 101, Room A903
Gaithersburg, MD 20899
(301) 975-2758

National Park Service
1849 C Street, NW
Washington, DC 20240
(202) 208-7394

National Transportation Safety Board
490 L'Enfant Plaza East, SW
Washington, DC 20594
(202) 382-6717

Nuclear Regulatory Commission
1555 Rockville Pike
Rockville, MD 20852
(301) 415-7000

Occupational Safety and Health Review
1120 20th Street, NW
Washington, DC 20036-3419
(202) 606-5100

Office of Science and Technology Policy
Old Executive Office Building, Room 435
17th and Pennsylvania Avenue, NW
Washington, DC 20502
(202) 456-6100

Peace Corps
1990 K Street, NW
Washington, DC 20526
(202) 606-3886

Smithsonian Institution
900 Jefferson Drive, SW
Washington, DC 20560
(202) 357-2627

US Fish and Wildlife Service
1849 C Street, NW
Washington, DC 20240
(202) 208-5634

Walter Reed Medical Hospital
6900 Georgia Avenue, NW
Washington, DC 20307
(202) 782-3501

Resources for Opportunities with the Government

The following resources provide more specific information on employment opportunities within local and national governments:

101 Challenging Government Jobs for College Graduates by William Shanahan (1986). Prentice Hall Press.

Almanac of American Government Jobs and Careers by Ronald Krannich (1991). Impact Publications (703) 361-7300.

Applying for Federal Jobs: A Guide to Writing Successful Applications and Resumes for the Job You Want in Government by Patricia Wood (1995). Brookhaven Press.

The Book of States, 1995-96 by The Council of State Governments (Lexington, KY). (800) 800-1910.

The Book of US Government Jobs: Where They Are, What's Available, and How to Get One by Dennis Damp (1994). D-Amp Publications.

Capitol Jobs: An Insider's Guide to Finding a Job in Congress by Kerry Dumbaugh and Gary Serota. Tilden Press (202) 332-1700.

Career Choices for the 90s: For Students of Political Science and Government (1990). Walker Publishers.

Careers in State and Local Government by John William Zehring. Garret Park Press (301) 946-2553.

The Complete Guide to Public Employment: Opportunities and Strategies with Federal, State, and Local Governments by Ronald Krannich and Caryl Rae Krannich (1995). Impact Publications (703) 361-7300.

Complete Guide to US Civil Service Jobs (1991). Acro Publishers.

The Congressional Intern Handbook by Sue Grabowski (1996). Congressional Management Foundation (202) 546-0100.

CRS Report for Congress—Internships and Fellowships: Congressional, Federal, and Other Work Experience Opportunities by Betsy Reifsnyder. Congressional Research Service in Washington, DC. Contact your Member of Congress and ask for a copy. Capitol Hill switchboard (202) 224-3121.

Federal Career Opportunities (a biweekly newsletter containing articles on issues relating to finding and applying for jobs in the Federal Government). Federal Research Service (703) 281-0200.

Find a Federal Job Fast!! by Ronald Krannich (1992). Impact Publications (703) 361-7300.

The Government Job Finder by Daniel Lauber (1995). Career Communications (800) 346-1848.

How to Get a Federal Job by David Waelde (1989). FEDHELP Publications.

Opportunities in State and Local Government Careers by Neale Baxter (1993). VGM Career Horizons (800) 323-4900.

Storming Washington: An Intern's Guide to National Government (5th edition) by Stephen Frantzich (1994). American Political Science Association (202) 483-2512.

The United States Government Manual 1996-97 from the Office of the Federal Register, National Archives and Records Administration, Government Printing Office. (202) 512-1800.

US Government Jobs: Career Opportunities for College Students (1993). Baron's Educational Series.

International Opportunities

The opportunity to work in another country can change your perspective and open up a plethora of new career options. While modern technology has truly transformed this planet into a global community, the experience of working within another culture will help you grow personally and professionally. In the 1990s, many organizations and business have an interest in the global aspects of the issues on which they work.

The organizations listed below offer opportunities in an international setting or they work directly with international issues. They offer a number of opportunities for young people to have the adventure of a lifetime and a chance to develop new skills through work in another country and culture. If travel to another country is not an option for you, but you are interested in international affairs or policy, these organizations may have useful information about how to work on international issues in the US. While this list is far from comprehensive, it should help you get started on your search.

Accion International
120 Beacon Street
Somerville, MA 02143
(617) 492-4930

Amigos de las Americas
5618 Star Lane
Houston, TX 77057
(800) 231-7796

Amnesty International
1118 22nd Street, NW
Washington, DC 20037
(202) 775-5161

Bretheren Volunteer Service
1451 Dundee Avenue
Elgin, IL 60120
(800) 323-8039

Carnegie Council on Ethics and International Affairs
170 E. 64th Street
New York, NY 10021-7478
(212) 838-4120

Citizens Democracy Corps
1400 I Street, NW, Suite 1125
Washington, DC 20005
(202) 872-0933

Institute for International Cooperation and Development
1117 Hancock Road
Williamstown, MA 01267
(413) 458-9828

International Development Exchange (IDEX)
827 Valencia Street, Suite 101
San Francisco, CA 94110
(415) 824-8384

International Institute for Energy Conservation
750 First Street, NE, Suite 940
Washington, DC 20002
(202) 842-3388

International Voluntary Services
1000 Connecticut Avenue, NW, Suite 901
Washington, DC 20036
(202) 387-5533

Los Niños
287 G Street
Chula Vista, CA 91910
(619) 426-9110

Civil Service International Workcamps
Innisfree Village,
Rt. 2, Box 506
Crozet, VA 22932

Concern/America
2024 N. Broadway, Suite 104
PO Box 1790
Santa Ana, CA 92706
(714) 953-8575

Conservation International
1015 18th Street, NW, Suite 1000
Washington, DC 20036
(202) 429-5660

Crossroads Africa
475 Riverside Drive, Suite 242
New York, NY 10115
(212) 870-2106

Earthwatch
680 Mt. Auburn Street
PO Box 9104
Box 403
Watertown, MA 02272-9104
(800) 776-0188

Finance for Development, Inc.
1730 Rhode Island Avenue, NW, Suite 610
Washington, DC 20036
(202) 467-0881

Global Citizens Network
1931 Iglehart Avenue
St. Paul, MN 55104
(612) 644-0960

Global Exchange
2017 Mission Street, #303
San Francisco, CA 94110
(415) 255-7296

Global Exchange/Study Associates International, Inc.
Box 661-College Station
New York, NY 10030-9998
(212) 781-4886

Global Volunteers
375 E. Little Canada Road
St. Paul, MN 55117
(612) 482-1074

National Council for International Health
1701 K Street, NW, Suite 600
Washington, DC 20006
(202) 833-5900

National Peace Foundation
1835 K Street, NW, Suite 610
Washington, DC 20006
(202) 223-1170

Overseas Development Network
333 Valencia Street, Suite 330
San Francisco, CA 94103
(415) 431-4204

Peace Corps
Box 718
Washington, DC 20526
(202) 606-3000 x 718

Peacework
305 Washington Street, NW
Blacksburg, VA 24060
(540) 953-1376

Refugees International
21 Dupont Circle
Washington, DC 20036
(202) 828-0110

Southern Center for International Studies
320 W. Paces Ferry Road, NW
Atlanta, GA 30305
(404) 261-5763

Taking Off
PO Box 104
Newton, MA 02161
(617) 630-1606

University Research Expeditions Program
University of California Berkeley
2223 Fulton Street, 4th Floor
Berkeley, CA 94720-7050
(510) 642-6586

Volunteers for Peace
43 Tiffany Road
Belmont, VT 05730
(802) 259-2759

Habitat for Humanity International
121 Habitat Street
Americus, GA 31709
(912) 924-6935 x124

International Christian Youth Exchange
134 W. 26th Street
New York, NY 10001
(212) 206-7307

World Federalist Association
418 7th Street, SE
Washington, DC 20003
(202) 546-3950

WorldTeach
Harvard Institute for International Development
One Eliot Street
Cambridge, MA 02138-5705
(617) 495-5527

Resources for International Opportunities

The following resources may help you find a job with international connections, both in the United States and abroad:

The Almanac of International Jobs and Careers: A Guide to Over 1001 Employers by Robert Krannich (1994). Impact Publishers (703) 361-7300.

The Complete Guide to International Jobs and Careers by Ronald Krannich (1992). Impact Publications (703) 361-7300.

Computer Jobs Worldwide: New-Authentic Jobs (1995). Zinks Publisher.

Guide to Careers in World Affairs by Foreign Policy Association (1993). Impact Publishers (703) 361-7300.

International Jobs: Where They are and How to Get Them by Eric Kocher (1993). Addison-Wesley Publishers.

Jobs for People Who Love Travel by Drs. Ron and Caryl Krannich (1995). Impact Publishers (703) 361-7300.

Jobs in Russia and the NIS by Moria Forbes (1994). Impact Publishers (703) 361-7300.

Jobs Worldwide by David Lay and Benedict A. Leerburger (1996). Impact Publishers (703) 361-7300.

Teaching English Abroad by Susan Griffith (1994). Impact Publishers (703) 361-7300.

Researching Corporate Responsibility

You've just been offered a great job—in your chosen field, perfectly fitting your skills, offering a great salary and benefits package—for a private sector company. What can a socially-minded person like yourself do to check out the social justice and environmental responsibility practices of a business? Today quite a few profit-making institutions practice environmental and social accountability.

As you begin to think about working for various corporations (or government agencies or nonprofit organizations), you should try to figure out which issues are of greatest importance to you. On which issues are you willing to compromise? There may be a few key points you want to look for in a company—its philanthropic practices, representation of women and minorities in senior management or on the board of directors, community service projects and volunteer programs for its employees, involvement in the weapons industry, or environmental practices.

Resources on this subject can be found in libraries and bookstores. In addition, the companies should have information available. The following list provides contact information for organizations that would likely have literature or know where to find the information you need to evaluate corporate responsibility.

Business Ethics
52 S.10th Street, Suite 110
Minneapolis, MN 55403-2001
(612) 962-4700

Business Ethics Strategies, Inc. (BEST)
PO Box 1698
New York, NY 10011
(212) 691-1224

Businesses for Social Responsibility
1030 15th Street, NW, Suite 1010
Washington, DC 20005
(202) 842-5400

Center for Business Ethics
Bentley College
175 Forest Street
Waltham, MA 02154
(617) 891-2981

Center for Corporate Public Involvement
1001 Pennsylvania Avenue, NW
Washington, DC 20004-2599
(202) 624-2425

Interfaith Center on Corporate Responsibility
475 Riverside Drive, Room 566
New York, NY 10115-0500
(212) 870-2295

International Association of Business and Society
c/o Sage Publications, Inc.
2455 Teller Road
Thousand Oaks, CA 91320
(805) 499-0721

The Josephson Institute of Ethics
4640 Admirality Way, Suite 1001
Marina Del Rey, CA 90292-6610
(310) 306-1868

National Network for Minority Women in Science
1333 H Street, NW, Room 1130
Washington, DC 20005
(202) 326-6757

Center for Professional Ethics
Manhattan College
Bronx, NY 10471
(718) 920-0442

**Center for the Study of Ethics
in the Professions**
Illinois Institute of Technology
3101 S. Dearborn Street
Room 166, Life Sciences Building
Chicago, IL 60616-3793
(312) 567-3017

Citizens Action
1730 Rhode Island Avenue, NW
Washington, DC 20036
(202) 775-1580

**Computer Professionals for Social
Responsibility**
PO Box 717
Palo Alto, CA 94302
(415) 322-3778

Council of Ethical Organizations
1216 King Street, Suite 300
Alexandria, VA 22314
(703) 683-7916

**Ethics Institute/The Right Thing, Inc.
(TRT)**
1106 Wilson Blvd., Suite 1711
Arlington, VA 20209
(703) 807-1162

Forum for Corporate Responsibility
593 Park Avenue
New York, NY 10021
(212) 758-2245

Institute of Business Ethics
301 East 47th Street, Suite 20-M
New York, NY 10017
(212) 832-8348

**Institute of International
Business Ethics**
15249 N. 59th Avenue
Glendale, AZ 85306-6000
(602) 978-7011

Renew America
1400 16th Street, NW, Suite 710
Washington, DC 20036
(202) 232-2252

Students for Responsible Business
1388 Sutter Street, Suite 1010
San Francisco, CA 94109
(415) 771-4308

Section 4
MORE ABOUT STUDENT PUGWASH USA

Student Pugwash USA: Who We Are . . .

The mission of Student Pugwash USA is to promote the socially responsible application of science and technology in the 21st century. As a student organization, Student Pugwash USA encourages young people to examine the ethical, social, and global implications of science and technology, and to make these concerns a guiding focus of their academic and professional endeavors.

Student Pugwash USA was founded in 1979. We are modeled after the 1995 Nobel Peace Prize recipients, the Pugwash Conferences on Science and World Affairs. With the advent of the hydrogen bomb as a humbling and frightening backdrop, Albert Einstein, Bertrand Russell, Joseph Rotblat, and others signed a manifesto urging scientists to consider the social, moral, and ethical implications of weapons of mass destruction. This manifesto led to the first Pugwash Conference, held in Pugwash, Nova Scotia in 1957. The Pugwash spirit has always implied the need for scientists to broadly consider the ethical implications of their work, beyond the challenges raised by nuclear weapons. At Student Pugwash USA, we draw on the Pugwash tradition as we seek to prepare young people to meet and to predict emerging challenges posed by science and technology. Student Pugwash USA is one of over 20 national Student and Young Pugwash groups around the world.

To accomplish its mission, Student Pugwash USA coordinates a wide range of programs and produces issue-oriented and leadership resources. Recognizing the need for fresh perspectives in today's debates, Student Pugwash USA's programs strive to bridge traditional academic disciplines and enhance an understanding of the interplay of scientific advancement, technological development, and the creation of an ethical society. Student Pugwash USA's initiatives are designed to involve students and experts in a way that is educational, inspirational, and fun.

. . . and What We Do

International Conferences: Student Pugwash USA's biennial international conferences draw upon the energy and knowledge of student and expert participants from countries around the world. The most recent conference addressed "Science, Technology, and Ethical Priorities" and tackled the following subjects: access and the Internet, alternative energy sources, emerging infectious diseases, international weapons trade, public participation in scientific decision making, and water quality and availability.

National Conferences: Student Pugwash USA sponsors annual national conferences, providing an opportunity for students from across the country to discuss issues in science and technology and to sharpen their organizing and leadership skills.

Regional Events: Regional events provide students with a forum to discuss issues with experts in various fields, meet with industry and political leaders, and network with other students and young profesionals. These are times when past and current members can get together.

Chapter Activities: Student Pugwash USA chapters creatively inject ethical and social dimensions into examinations of science and technology on campuses across the United States.

Mentorship: Student Pugwash USA's Alternative Mentorship Network (AMNet) provides an opportunity for students to develop individualized relationships with professionals who can offer advice on graduate study, research topics, and career development.

Publications: In adition to *Jobs You Can Live With*, Student Pugwash USA produces:
Tough Questions—our biannual newsletter on critical global and social issues.
Chapter Organizing Guide—a comprehensive resource manual for students organizing local and regional activities.
Pugwatch—a monthly news bulletin for student members of the chapter program.
Mind•full: A brainsnack for future leaders with ethical appetites—issue briefs which provide an overview and insight to key global concerns.

Section 5
INDICES

Organizations At-A-Glance

Organizations	Energy	Environment	Food & Agriculture	Global Security	Health & Bio-med. Research	Info. Tech.
20/20 Vision		●		●		
Access: A Security Information Service				●		
ACCESS: Networking in the Public Interest						●
ACCION International						●
Advocacy Institute	●	●	●	●	●	●
African Scientific Institute	●	●	●	●	●	●
Alaska Green Goods	●	●				
Alice Hamilton Occupational Health Center		●				●
Alternative Energy Resources Organization (AERO)	●	●	●			
American Forests		●				
American Hydrogen Association	●	●				
Antarctica Project		●				
Appalachian Mountain Club		●				
Arizona Toxics Information		●				
Arms Control Association				●		
Ashoka: Innovators for the Public Interest		●			●	●
Association for Women in Sciences (AWIS)						●
Association to Unite the Democracies		●		●		
Atlantic Council of the United States	●	●		●		
Benton Foundation					●	●
Bering Sea Coalition	●	●	●	●	●	●
Bigelow Laboratory for Ocean Sciences		●				
Blue Ridge Environmental Defense League, Inc.	●	●	●			

Organizational Description

Educa-tional	Lobbies	Works with Media	Public Policy & Admin.	Commun. Organiz-ing	Uses Econ. Analysis	Address Gender Issues	Address Racial Issues	Address Ethics	Address Legal Issues	Inter-national focus	Science Tech. & Society
●	●	●	●	●							
●		●	●							●	
●			●	●							
			●	●		●	●	●		●	
●			●	●	●		●	●	●		●
●				●						●	
●								●			
●								●	●		
●		●		●							
●		●		●						●	
●										●	
●				●					●	●	
●				●							
●		●	●							●	●
●		●								●	●	●
●		●		●		●	●	●	●	●	
●	●			●		●	●			●	
●	●	●	●							●	
●		●	●							●	
		●	●	●						●	
●	●	●	●	●	●	●	●	●		●	
●	●	●	●	●						●	
●		●		●				●	●	●	

Organizations At-A-Glance

Subject Area

Organization	Energy	Environment	Food & Agriculture	Global Security	Health & Bio-med. Research	Info. Tech.
Bread for the World			●			
Break Away: The Alternative Break Connection		●	●		●	
British American Security Council (BASIC)				●		
Campaign for UN Reform		●		●		●
Campus Green Vote		●				●
Campus Outreach Opportunity League (COOL)		●				●
Career Development Group (CDG)						●
Carrying Capacity Network		●	●			
Carter Center	●	●	●	●	●	
Center for Campus Organizing						●
Center for Clean Air Policy	●	●				
Center for Defense Information (CDI)				●		
Center for Democracy and Technology						●
Center for Economic Conversion (CEC)	●	●		●		
Center for Global Change	●	●				
Center for Integrated Agricultural Systems		●	●			
Center for Marine Conservation		●				
Center for Policy Alternatives		●				
Center for Psychology and Social Change		●				
Center for Rural Affairs		●	●			
Center for Science in the Public Interest		●	●		●	
Center for the Study of Ethics in the Professions (CSEP)					●	●

Organizational Description

Educational	Lobbies	Works with Media	Public Policy & Admin.	Commun. Organizing	Uses Econ. Analysis	Address Gender Issues	Address Racial Issues	Address Ethics	Address Legal Issues	International focus	Science Tech. & Society
●	●	●	●	●				●		●	
●		●	●	●							
		●	●							●	
	●	●	●		●			●	●	●	
●		●		●				●			
●		●	●	●		●	●	●			
●		●	●								●
●		●			●						
●	●	●		●	●		●	●		●	
●		●		●	●	●	●		●	●	
		●	●		●						
	●	●	●	●							
●	●		●							●	
●		●		●	●						
●			●		●				●	●	
●		●		●		●	●	●	●	●	
●		●		●						●	
●				●		●	●	●	●	●	
				●		●	●			●	
●	●	●	●	●	●			●	●		
●	●	●	●	●						●	
●		●						●	●		

265

Organizations At-A-Glance

Subject Area

Organizations	Energy	Environment	Food & Agriculture	Global Security	Health & Bio-med. Research	Info. Tech.
Center for War, Peace and the News Media				●		
Center for Women Policy Studies					●	
Chesapeake Bay Foundation		●				
Citizens Clearinghouse for Hazardous Waste (CCHW)		●				
Citizens for Alternatives to Chemical Contamination (CACC)		●				
Clean Water Action		●				
Co-op America	●	●				
Committee for the National Institute for the Environment (CNIE)		●				
Common Cause				●		
CONCERN, Inc.	●	●				
Concord Feminist Health Center (CFHC)					●	
Conservation Career Development Program (CCDP)		●				
Conservation International		●				
Consumer Energy Council of America Research Foundation (CECA/RF)	●	●				
Consanti Foundation	●	●				
Council For A Livable World				●		
Council for Responsible Genetics			●		●	
Council on Economic Priorities			●	●		
Council on Hemispheric Affairs (COHA)	●	●	●	●		
Cultural Survival, Inc.				●		
Dakota Resource Council		●	●			

Organizational Description

Educational	Lobbies	Works with Media	Public Policy & Admin.	Commun. Organizing	Uses Econ. Analysis	Address Gender Issues	Address Racial Issues	Address Ethics	Address Legal Issues	International focus	Science Tech. & Society
		•	•					•	•	•	
•		•	•	•		•	•	•			
•	•	•	•							•	
•				•		•	•	•	•		
•		•	•	•							
•	•	•	•	•							
•		•		•							
•	•	•	•	•							
	•	•	•								
•				•							
•	•	•		•		•	•	•	•		
•			•	•		•	•				
		•		•	•					•	
			•	•	•						
•		•		•				•		•	
•	•	•	•							•	
•		•	•					•	•		•
					•	•	•	•		•	
•		•	•		•	•	•	•	•	•	
•			•		•	•	•			•	
•	•	•	•	•							

267

Organizations At-A-Glance

Subject Area

Organization	Energy	Environment	Food & Agriculture	Global Security	Health & Bio-med. Research	Info. Tech.
Dakota Rural Action (DRA)		●	●			
Defenders of Wildlife		●				
Delaware Valley Citizens' Council For Clean Air (Clean Air Council)		●				
Demilitarization for Democracy (DFD)				●		
Dodge Nature Center		●	●			
E: The Environmental Magazine	●	●				
Earth Island Institute		●				
Earth Share		●				
Earth Train		●				
Earthtrust		●				
Ecological Society of America		●	●			
Educational Foundation for Nuclear Science				●		
Educators for Social Responsibility		●		●		
Energy and Environmental Concepts, Inc.	●	●				
Environmental Action Foundation	●	●				
Environmental Advantage	●	●				
Environmental Law Institute		●				
Environmental Working Group		●	●			●
Ethics and Public Policy Center		●			●	
Farm Sanctuary		●			●	
Federation of American Scientists				●		
Food and Water, Inc.		●	●			

Organizational Description

Educational	Lobbies	Works with Media	Public Policy & Admin.	Commun. Organizing	Uses Econ. Analysis	Address Gender Issues	Address Racial Issues	Address Ethics	Address Legal Issues	International focus	Science Tech. & Society
●	●	●	●	●	●			●	●		
●	●	●	●	●					●		
●		●	●	●					●		
	●	●	●				●	●	●	●	
●				●				●			
●		●									
●		●	●	●			●	●	●	●	
●		●	●	●							
●		●		●						●	
	●	●	●					●	●	●	
●		●	●								
●		●						●		●	
●		●				●	●	●		●	●
●		●									
●				●							
					●					●	
●			●		●				●		
●		●	●	●	●				●		
●		●	●					●	●	●	
●		●						●	●		
	●	●		●				●			
●		●		●				●			

Organizations At-A-Glance

Subject Area

	Energy	Environment	Food & Agriculture	Global Security	Health & Bio-med. Research	Info. Tech.
Friends Committee on National Legislation (FCNL)				●		
Friends of the Boundary Water Wilderness		●				
Friends of the Earth	●	●				
Friends of the River		●				
Fund for Animals, Inc.		●				
Global Change	●	●				
Global Exchange			●	●		
Government Accountability Project	●	●	●			
Greater Yellowstone Coalition		●				
Green Corps	●	●				
Greenlining Institute						●
Greenwire		●				
Habitat for Humanity International					●	
HawkWatch International		●				
Heal the Bay		●				
Herbert Scoville Jr. Peace Fellowship				●		
Hispanic Association of Colleges and Universities (HACU)	●	●	●	●	●	●
Human Rights Watch (HRW)				●		
Information Data Management					●	●
Innovative System Developers, Inc. (ISD)		●	●		●	●
Institute for Development Anthropology (IDA)		●	●			
Institute for Global Communications (IGC)		●		●		●

Organizational Description

Educational	Lobbies	Works with Media	Public Policy & Admin.	Commun. Organizing	Uses Econ. Analysis	Address Gender Issues	Address Racial Issues	Address Ethics	Address Legal Issues	International focus	Science Tech. & Society
●	●			●	●			●			
●	●			●				●			
●	●	●	●	●	●	●		●	●	●	
●	●		●	●							
●	●	●	●		●			●	●		
●		●								●	
●				●	●		●			●	
●		●	●					●	●		
●	●	●		●	●				●		
●	●	●		●							
●	●	●	●	●	●	●	●	●			
		●	●							●	
		●	●	●	●					●	
●		●	●	●						●	
●	●	●	●	●	●			●	●		
●			●					●		●	
●				●			●				
	●	●	●			●	●		●	●	
●					●				●	●	
●			●		●						
●		●	●	●	●	●	●	●	●	●	
●					●	●	●	●	●		

Organizations At-A-Glance

Subject Area

	Energy	Environment	Food & Agriculture	Global Security	Health & Bio-med. Research	Info. Tech.
Institute for Global Ethics		●		●		
Institute for International Cooperation and Development (IICD)		●	●			
Institute for Local Self-Reliance	●	●	●			
Institute for Mental Health Initiatives (IMHI)					●	
Institute for Peace and Justice				●		
Institute for Policy Studies			●	●	●	
Institute for Resource and Security Studies	●	●		●		
Institute for Science and International Security (ISIS)				●		
Institute of World Affairs		●		●		
International Center for Research on Women (ICRW)		●	●		●	
International Development and Energy Associates, Inc. (IDEA)	●	●				
International Development Exchange (IDEX)			●			
International Institute for Energy Conservation (IIEC)	●	●				
Issues in Science and Technology	●	●	●	●	●	●
Izaak Walton League of America, Inc.	●	●				
Just Harvest: A Center for Action Against Hunger			●		●	
Land Institute			●			
Land Stewardship Project (LSP)		●	●			
Lawyers Alliance for World Security				●		
Lawyers Committee on Nuclear Policy				●		

Organizational Description

Educational	Lobbies	Works with Media	Public Policy & Admin.	Commun. Organizing	Uses Econ. Analysis	Address Gender Issues	Address Racial Issues	Address Ethics	Address Legal Issues	International focus	Science Tech. & Society
●								●		●	●
●										●	
●			●		●						
●		●		●		●	●				
●	●					●	●	●		●	
●		●	●		●	●	●	●		●	
●		●	●		●			●	●	●	
●		●	●							●	
●				●	●			●		●	
●		●	●			●	●			●	
●			●						●	●	
●				●			●			●	
			●		●					●	
●			●					●	●		
●			●					●			
●	●	●	●	●	●			●			
●			●					●			
●	●		●	●				●			
●		●	●							●	
●			●	●				●	●	●	

273

Organizations At-A-Glance

Subject Area

Organization	Energy	Environment	Food & Agriculture	Global Security	Health & Bio-med. Research	Info. Tech.
League of Conservation Voters (LCV)		●				
Leveraged Outreach Project				●		
Living Classrooms Foundation		●				
Loka Institute						●
Los Niños		●	●		●	
Louisiana Nature Center	●	●	●	●	●	●
Maine Organic Farmers & Gardeners Association (MOFGA)	●	●				
Management Sciences for Health (MSH)					●	●
Matsunaga Institute for Peace at the University of Hawaii				●		
Meadowcreek	●	●	●			
Medical Technology and Practice Patterns Institute (MTPPI)					●	
Metasystems Design Group, Inc.						●
Mid-South Peace and Justice Center		●		●		
Midwest Renewable Energy Association	●					
National Action Council for Minorities in Engineering (NACME)						●
National Association for Science, Technology, and Society (NASTS)	●	●			●	●
National Audubon Society		●	●			
National Center for Appropriate Technology (NCAT)	●	●	●			●
National Coalition Against the Misuse of Pesticides (NCAMP)		●	●			
National Council for International Health				●	●	

Organizational Description

Educational	Lobbies	Works with Media	Public Policy & Admin.	Commun. Organizing	Uses Econ. Analysis	Address Gender Issues	Address Racial Issues	Address Ethics	Address Legal Issues	International focus	Science Tech. & Society
●	●	●	●	●							
●		●	●					●		●	
●											
●		●	●	●		●	●	●			●
●				●						●	
●				●							
●			●	●							
●			●	●	●	●				●	
●				●			●	●		●	
●				●				●			
			●		●					●	
●				●				●			
●	●	●		●				●		●	
●				●							
		●	●			●	●				
●						●		●		●	●
●	●	●	●	●							
●	●		●	●	●						
●		●	●	●			●	●	●		
●			●	●		●				●	

Organizations At-A-Glance

	Subject Area					
	Energy	Environment	Food & Agriculture	Global Security	Health & Bio-med. Research	Info. Tech.
National Network of Minority Women in Science (MWIS)	●	●			●	●
National Organization for Women (NOW)					●	
National Peace Foundation				●		
National Recycling Coalition (NRC)		●				
National Security and Natural Resources News Service		●		●		
National Security Archive				●		●
National Wildlife Federation		●				
National Women's Health Network					●	
Natural Resources Council of Maine		●				
Natural Resources Defense Council (NRDC)	●	●	●			
Nature Conservancy		●				
NGO Committee on Disarmament, Inc.				●		
Northern Alaska Environmental Center		●				
Nuclear Age Peace Foundation				●		
Overseas Development Network				●		
Pacific Whale Foundation (PWF)		●				
Palouse-Clearwater Environmental Institute		●	●			
Peace Action				●		
Peace and Justice Center				●		
Pesticide Action Network North America		●	●		●	
Physicians for Human Rights (PHR)				●	●	
Physicians for Social Responsibility (PSR)	●	●			●	

Organizational Description

Educational	Lobbies	Works with Media	Public Policy & Admin.	Commun. Organizing	Uses Econ. Analysis	Address Gender Issues	Address Racial Issues	Address Ethics	Address Legal Issues	International focus	Science Tech. & Society
●			●			●	●				
	●	●	●	●		●	●		●		
●	●									●	
●		●	●	●							
			●							●	
●		●							●	●	
●	●	●	●	●	●			●	●	●	
●	●		●	●		●		●	●		
●	●	●	●	●				●	●		
●	●		●	●						●	
●	●	●	●	●						●	
●								●		●	
●	●	●	●	●							
●	●			●				●	●	●	
●	●	●	●	●				●		●	
●		●		●						●	
●		●		●							●
●	●		●	●	●			●		●	
●			●	●			●				
●	●		●			●	●	●	●	●	
●			●	●					●	●	
●	●	●	●	●				●			

Organizations At-A-Glance

Organization	Energy	Environment	Food & Agriculture	Global Security	Health & Bio-med. Research	Info. Tech.
League of Conservation Voters (LCV)		●				
Leveraged Outreach Project				●		
Living Classrooms Foundation		●				
Loka Institute						●
Los Niños		●	●		●	
Louisiana Nature Center	●	●	●	●	●	●
Maine Organic Farmers & Gardeners Association (MOFGA)	●	●				
Management Sciences for Health (MSH)					●	●
Matsunaga Institute for Peace at the University of Hawaii				●		
Meadowcreek	●	●	●			
Medical Technology and Practice Patterns Institute (MTPPI)					●	
Metasystems Design Group, Inc.						●
Mid-South Peace and Justice Center		●		●		
Midwest Renewable Energy Association	●					
National Action Council for Minorities in Engineering (NACME)						●
National Association for Science, Technology, and Society (NASTS)	●	●			●	●
National Audubon Society		●	●			
National Center for Appropriate Technology (NCAT)	●	●	●			●
National Coalition Against the Misuse of Pesticides (NCAMP)		●	●			
National Council for International Health				●	●	

Organizational Description

Educational	Lobbies	Works with Media	Public Policy & Admin.	Commun. Organizing	Uses Econ. Analysis	Address Gender Issues	Address Racial Issues	Address Ethics	Address Legal Issues	International focus	Science Tech. & Society
●	●	●	●	●							
●			●	●				●		●	
●											
●			●	●	●		●	●	●			●
●				●						●	
●				●							
●			●	●							
●				●	●	●	●				●	
●				●			●	●		●	
●				●				●			
			●		●					●	
●				●				●			
●	●	●		●				●		●	
●				●							
		●	●				●	●			
●							●	●	●		●	●
●	●	●	●	●							
●	●			●	●	●					
●			●	●	●				●	●	●	
●			●	●			●				●	

Organizations At-A-Glance

Subject Area

Organization	Energy	Environment	Food & Agriculture	Global Security	Health & Bio-med. Research	Info. Tech.
National Network of Minority Women in Science (MWIS)	●	●			●	●
National Organization for Women (NOW)					●	
National Peace Foundation				●		
National Recycling Coalition (NRC)		●				
National Security and Natural Resources News Service		●		●		
National Security Archive				●		●
National Wildlife Federation		●				
National Women's Health Network					●	
Natural Resources Council of Maine		●				
Natural Resources Defense Council (NRDC)	●	●	●			
Nature Conservancy		●				
NGO Committee on Disarmament, Inc.				●		
Northern Alaska Environmental Center		●				
Nuclear Age Peace Foundation				●		
Overseas Development Network				●		
Pacific Whale Foundation (PWF)		●				
Palouse-Clearwater Environmental Institute		●	●			
Peace Action				●		
Peace and Justice Center				●		
Pesticide Action Network North America		●	●		●	
Physicians for Human Rights (PHR)				●	●	
Physicians for Social Responsibility (PSR)	●	●			●	

Organizational Description

Educational	Lobbies	Works with Media	Public Policy & Admin.	Commun. Organizing	Uses Econ. Analysis	Address Gender Issues	Address Racial Issues	Address Ethics	Address Legal Issues	International focus	Science Tech. & Society
●			●			●	●				
	●	●	●	●		●	●		●		
●	●									●	
●		●	●	●							
		●								●	
●		●							●	●	
●	●	●	●	●	●			●	●	●	
●	●		●	●			●	●	●		
●	●	●	●	●				●	●		
●	●		●	●						●	
●	●	●	●	●						●	
●								●		●	
●	●	●	●	●							
●	●		●					●	●	●	
●	●	●	●	●				●			
●		●		●						●	
●		●		●							●
●	●		●	●	●			●		●	
●			●	●			●				
●	●		●			●	●	●	●	●	
●			●	●				●		●	
●	●	●	●	●				●			

Organizations At-A-Glance

Subject Area

Organization	Energy	Environment	Food & Agriculture	Global Security	Health & Bio-med. Research	Info. Tech.
Plugged In		●				●
Population Association of America	●	●	●	●	●	●
Powder River Basin Resource Council	●	●	●			
Public Citizen Critical Mass Energy Project	●	●	●			
Public Voice for Food and Health Policy		●	●		●	
Rainforest Action Network		●				
Refugees International				●		
Renew America		●				
Rhode Island Solar Energy Association	●					●
River Watch Network (RWN)		●				
Riveredge Nature Center		●				
Rocky Mountain Institute	●	●	●	●		
Search for Common Ground				●		
Shaver's Creek Environmental Center		●				
Sigma Xi, the Scientific Research Society	●	●	●		●	
Smithsonian Environmental Research Center (SERC)		●				
Social Action and Leadership School for Activists						●
Society of Environmental Journalists		●				
Solar Energy Industries Association (SEIA)	●	●				
Southeast Alaska Conservation Council		●				
Southern Technology Council						●
Southface Energy Institute	●	●				
Southwest Network for Environmental and Economic Justice		●		●		

Organizational Description

Educa-tional	Lobbies	Works with Media	Public Policy & Admin.	Commun. Organiz-ing	Uses Econ. Analysis	Address Gender Issues	Address Racial Issues	Address Ethics	Address Legal Issues	Inter-national focus	Science Tech. & Society
●		●	●	●				●		●	
●	●	●	●	●	●	●	●	●	●	●	●
		●		●	●			●			
●	●		●		●				●		●
●	●	●	●		●						
●		●	●	●	●			●	●	●	
	●	●	●				●	●			
●		●	●	●							
●		●		●						●	
●				●						●	
●				●						●	
●		●	●	●						●	
●		●		●		●	●			●	
●				●							
●		●	●	●				●		●	
●										●	●
●			●	●				●			
●		●								●	
●	●	●								●	
●	●	●	●	●	●				●		
●			●		●						●
●		●		●	●						
●		●	●	●	●	●	●	●	●	●	

Organizations At-A-Glance

Organization	Energy	Environment	Food & Agriculture	Global Security	Health & Bio-med. Research	Info. Tech.
Southwest Research and Information Center (SRIC)		●				
Structural Dynamics Research Corporation (SDRC)						●
Student Conservation Association - Resource Assistant Program	●	●				
Student Conservation Association (SCA)		●				
Student Pugwash USA (SPUSA)	●	●	●	●	●	●
Taking Off	●	●	●	●	●	●
Teach for America						●
Tellus Institute, Inc.	●	●				
Tennessee Environmental Council		●				
Texas Environmental Center (TEC)		●				●
The Progressive Magazine	●	●		●		
Transportation Alternatives		●				
Traprock Peace Center		●		●		
Trees for the Future	●	●	●			●
Union of Concerned Scientists (UCS)	●	●		●		
University of Rhode Island Environmental Education Center		●	●			
Vermont Raptor Center		●				
Video Project	●	●	●	●		
Visions in Action		●	●		●	
Volunteers for Outdoor Colorado		●				
War and Peace Foundation				●	●	
War Resisters League (WRL)				●		

Organizational Description

Educational	Lobbies	Works with Media	Public Policy & Admin.	Commun. Organizing	Uses Econ. Analysis	Address Gender Issues	Address Racial Issues	Address Ethics	Address Legal Issues	International focus	Science Tech. & Society
●			●	●		●	●	●	●	●	
					●			●		●	
●				●							
●	●		●	●							
●								●		●	●
●	●	●	●	●	●	●	●	●	●	●	
●				●			●				
●		●	●		●					●	
●				●					●		
●		●	●					●			
●		●	●				●	●			
			●	●					●	●	
●			●	●				●			
●			●	●		●		●		●	
●	●		●					●		●	
●				●							
●			●					●			
●			●	●				●	●	●	
●			●	●						●	
●				●							
●	●	●	●	●				●	●	●	●
●				●			●	●	●		●

285

Organizations At-A-Glance

Subject Area

Organization	Energy	Environment	Food & Agriculture	Global Security	Health & Bio-med. Research	Info. Tech.
Washington Wilderness Coalition		●				
Wildlife Center of Virginia		●				
Wildlife Conservation Society		●				
William Penn House		●				
Women in International Security (WIIS)				●		
Women Strike for Peace	●	●		●		
Women's Action for New Directions				●		
World Federalist Association				●		
World Game Institute		●	●	●		●
World Service Authority		●		●		
Worldwatch Institute	●	●				●
YES! (Youth for Environmental Sanity)		●				
Zero Population Growth (ZPG)		●				

Organizational Description

Educational	Lobbies	Works with Media	Public Policy & Admin.	Commun. Organizing	Uses Econ. Analysis	Address Gender Issues	Address Racial Issues	Address Ethics	Address Legal Issues	International focus	Science Tech. & Society
●	●	●		●				●	●		
●		●	●					●	●		
●	●		●							●	
●								●		●	
●				●		●	●			●	
●	●	●		●						●	
●	●		●			●					
●	●			●					●	●	
●		●			●	●	●	●		●	
●			●	●		●	●	●		●	
●			●						●		●
●		●		●		●	●	●		●	
●	●	●		●						●	

Geographical Listing

Alaska
Alaska Green Goods
Bering Sea Coalition
Northern Alaska Environmental Center
Southeast Alaska Conservation Council

Arkansas
Meadowcreek

Arizona
American Hydrogen Association
Arizona Toxics Information
Cosanti Foundation

California
African Scientific Institute
Center for Economic Conversion (CEC)
Earth Island Institute
Earth Train
Friends of the River
Global Exchange
Greenlining Institute
Heal the Bay
Institute for Global Communications (IGC)
International Development Exchange (IDEX)
Los Niños
Nuclear Age Peace Foundation
Overseas Development Network
Pesticide Action Network North America
Plugged In
Rainforest Action Network
Video Project
YES! (Youth for Environmental Sanity)

Colorado
Rocky Mountain Institute
Volunteers for Outdoor Colorado

Connecticut
E: The Environmental Magazine

Georgia
Carter Center
Habitat for Humanity International
Southface Energy Institute

Hawaii
Earthtrust
Matsunaga Institute for Peace at the University of Hawaii
Pacific Whale Foundation (PWF)

Idaho
Palouse-Clearwater Environmental Institute

Illinois
Center for the Study of Ethics in the Professions (CSEP)
Educational Foundation for Nuclear Science
Information Data Management

Kansas
Land Institute

Louisiana
Louisiana Nature Center

Massachusetts
ACCION International
Appalachian Mountain Club
Center for Campus Organizing
Center for Psychology and Social Change
Council for Responsible Genetics
Cultural Survival Inc.
Educators for Social Responsibility
Green Corps
Institute for International Cooperation and Development (IICD)
Institute for Resource and Security Studies
Loka Institute
Management Sciences for Health (MSH)
Physicians for Human Rights (PHR)
Taking Off
Tellus Institute, Inc.
Traprock Peace Center
Union of Concerned Scientists (UCS)
Women's Action For New Directions

Maryland
Bread for the World
Center for Global Change
Chesapeake Bay Foundation
Environmental Action Foundation
Fund for Animals, Inc.
Innovative System Developers Inc. (ISD)
Izaak Walton League of America, Inc.
Living Classrooms Foundation
Population Association of America
Smithsonian Environmental
 Research Center (SERC)
Trees for the Future
Women in International Security (WIIS)

Maine
Bigelow Laboratory for Ocean Sciences
Institute for Global Ethics
Maine Organic Farmers & Gardeners
 Association (MOFGA)
Natural Resources Council of Maine

Michigan
Citizens for Alternatives to Chemical
 Contamination (CACC)
Energy and Environmental Concepts, Inc.

Minnesota
Dodge Nature Center
Friends of the Boundary Water Wilderness
Land Stewardship Project (LSP)

Missouri
Institute for Peace and Justice

Montana
Alternative Energy Resources
 Organization (AERO)
Greater Yellowstone Coalition
National Center for Appropriate
 Technology (NCAT)

North Carolina
Blue Ridge Environmental Defense
 League, Inc.
Sigma Xi, the Scientific Research Society
Southern Technology Council

North Dakota
Dakota Resource Council

Nebraska
Center for Rural Affairs

New Hampshire
Concord Feminist Health Center (CFHC)
Student Conservation Association—
 Resource Assistant Program
Student Conservation Association (SCA)

New Mexico
Southwest Network for Environmental
 and Economic Justice
Southwest Research and Information
 Center (SRIC)

New York
Career Development Group (CDG)
Center for War, Peace and the
 News Media
Council on Economic Priorities
Environmental Advantage
Farm Sanctuary
Institute for Development Anthropology
 (IDA)
Lawyers Committee on Nuclear Policy
National Action Council for Minorities in
 Engineering (NACME)
NGO Committee on Disarmament, Inc.
Teach for America
Transportation Alternatives
War and Peace Foundation
War Resisters League (WRL)
Wildlife Conservation Society

Geographical Listing (cont.)

Ohio
Structural Dynamics Research Corporation (SDRC)

Pennsylvania
Delaware Valley Citizens' Council For Clean Air (Clean Air Council)
Just Harvest: A Center for Action Against Hunger
National Association for Science, Technology, and Society (NASTS)
Shaver's Creek Environmental Center
Society of Environmental Journalists
World Game Institute

Rhode Island
Rhode Island Solar Energy Association
University of Rhode Island Environmental Education Center

South Dakota
Dakota Rural Action (DRA)

Tennessee
Break Away: The Alternative Break Connection
Mid-South Peace and Justice Center
Tennessee Environmental Council

Texas
Issues in Science and Technology
Texas Environmental Center (TEC)

Utah
HawkWatch International

Virginia
Ashoka: Innovators for the Public Interest
Citizens Clearinghouse for Hazardous Waste (CCHW)
Conservation Career Development Program (CCDP)
Greenwire
Metasystems Design Group Inc.
National Recycling Coalition (NRC)
Nature Conservancy
Wildlife Center of Virginia

Vermont
Food and Water, Inc.
Peace and Justice Center
River Watch Network (RWN)
Vermont Raptor Center

Washington
Washington Wilderness Coalition

Wisconsin
Center for Integrated Agricultural Systems
Midwest Renewable Energy Association
Riveredge Nature Center
The Progressive Magazine

Wyoming
Powder River Basin Resource Council

Washington, DC

20/20 Vision
Access: A Security Information Service
ACCESS: Networking in the
    Public Interest
Advocacy Institute
Alice Hamilton Occupational
    Health Center
American Forests
Antarctica Project
Arms Control Association
Association for Women in Science (AWIS)
Association to Unite the Democracies
Atlantic Council of the United States
Benton Foundation
British American Security Information
    Council (BASIC)
Campaign for UN Reform
Campus Green Vote
Campus Outreach Opportunity
    League (COOL)
Carrying Capacity Network
Center for Clean Air Policy
Center for Defense Information (CDI)
Center for Democracy and Technology
Center for Marine Conservation
Center for Policy Alternatives
Center for Science in the Public Interest
Center for Women Policy Studies
Clean Water Action
Co-op America
Committee for the National Institute for
    the Environment (CNIE)
Common Cause
CONCERN, Inc.
Conservation International
Consumer Energy Council of America
    Research Foundation (CECA/RF)
Council For A Livable World
Council on Hemispheric Affairs (COHA)
Defenders of Wildlife
Demilitarization for Democracy (DFD)
Earth Share
Ecological Society of America
Environmental Law Institute
Environmental Working Group
Ethics and Public Policy Center
Federation of American Scientists
Friends Committee on National
    Legislation (FCNL)
Friends of the Earth
Global Change
Government Accountability Project
Herbert Scoville Jr. Peace Fellowship
Hispanic Association of Colleges and
    Universities (HACU)
Human Rights Watch (HRW)
Institute for Local Self-Reliance
Institute for Mental Health Initiatives
    (IMHI)
Institute for Policy Studies
Institute for Science and International
    Security (ISIS)
Institute of World Affairs
International Center for Research on
    Women (ICRW)
International Development and Energy
    Associates, Inc. (IDEA)
International Institute for Energy
    Conservation (IIEC)
Lawyers Alliance for World Security
League of Conservation Voters (LCV)
Leveraged Outreach Project
Medical Technology and Practice Patterns
    Institute (MTPPI)
National Audubon Society
National Coalition Against the Misuse of
    Pesticides (NCAMP)
National Council for International Health
National Network of Minority Women in
    Science (MWIS)
National Organization for Women (NOW)
National Peace Foundation
National Security and Natural Resources
    News Service
National Security Archive
National Wildlife Federation
National Women's Health Network
Natural Resources Defense Council
    (NRDC)
Peace Action
Physicians for Social Responsibility (PSR)

Geographical Listing (cont.)

Washington, DC (cont.)
Public Citizen Critical Mass Energy Project
Public Voice for Food and Health Policy
Refugees International
Renew America
Search for Common Ground
Social Action and Leadership School
    for Activists
Solar Energy Industries Association (SEIA)
Student Pugwash USA (SPUSA)
Visions in Action
William Penn House
Women Strike for Peace
World Federalist Association
World Service Authority
Worldwatch Institute
Zero Population Growth (ZPG)